AN OVERVIEW OF
CORROSION, INHIBITORS
AND
JOURNALS

Dr Benjamin Valdez Salas PhD Nelson Cheng PhD (HC), SRF Patrick Moe BSc, MSc, Grad. Dip

To order additional copies of this book, contact
Toll Free +65 3165 7531 (Singapore)
Toll Free +60 3 3099 4412 (Malaysia)
www.partridgepublishing.com/singapore
orders.singapore@partridgepublishing.com

ISBN: 978-1-5437-7294-4 (sc)
ISBN: 978-1-5437-7296-8 (hc)
ISBN: 978-1-5437-7295-1 (e)

Print information available on the last page.

03/20/2023

PARTRIDGE

Acknowledgements

In gratitude for the help we received in publishing this book, we would like to thank God. We are thankful to all the authors for their valued contribution to this book.

The following people and institutions deserve special recognition:

Universidad Autónoma de Baja California (UABC), Mexicali, Baja California, Mexico

Benjamín Valdez Salas, Michael Schorr, Ernesto Beltran, Ricardo Salinas, Francisco Flores, R. Garcia, Roumen Zlatev Koytchev, Rogelio Ramos Irigoyen, Monica Carrillo Beltrán, Nicola Radnev Nedev, Mario Curiel Alvarez, J. S. Salvador-Carlos, R. G. Inzunza, A. Calderas, N. Santillan, and Jorge Ramirez.

Polytechnic University of Baja California, Calle de la Claridad S/N, Colonia Plutarco Elias Calles, 21376 Mexicali, BCN, Mexico Gustavo López Badilla

Navor Rosas Gonzalez

Proyecto Tropicorr—CYTED, CINVESTAV-IPN, Merida, Appl. Physics Dept., Carr. Ant. A Progreso, Km. 6, CP 97310, Merida, Yucatan, Mexico

Lucien Veleva

National Centre for Metallurgical Research, CSIC, Spain

Jose María Bastidas

Sami Shamoon College of Engineering—Beer-Sheeva, Israel

Amir Eliezer

Magna Chemical Canada Inc.

James Cheng

Our sincere thanks to the Patridge team for bringing this book to its final form.

Preface

According to NACE Impact Report dated March 2017, the global cost of corrosion is estimated to be US$2,505 billion (22 March 2017), which is equivalent to 3.4 per cent of the global GDP (2013). In addition, these costs typically do not include individual safety or environmental consequences.

This book covers an overview of the types of corrosion in each industry and the basics of corrosion inhibitors and their applications. It has been a challenge for mankind to deal with corrosion for ages, and it is one of the major catastrophes of the world.

It is extremely challenging and extremely important to protect metals from corrosion. Corrosion inhibitors are commonly used in various industries to combat corrosion. Chemicals such as these impede the corrosion of metals when added to corrosive media in traces. As they get adsorbed onto metal surfaces, they form a very thin film of molecules that protects them from further corrosion.

Corrosive environments and metal surfaces are retarded by this film. The reduction in corrosion can be attributed to the physical blockage effect, or to the inhibitor's influence on corrosion mechanisms and kinetics. These corrosion inhibitors are primarily found in solutions or dispersions, but some are also found in paints and coatings. When corrosion inhibitors are used, the severity of the corrosion problem can be reduced. Advances in corrosion control methods and techniques remain a major concern for industry and academia. The purpose of this chapter is to provide a brief overview of corrosion inhibitors.

In the twenty-first century, one of the most important global challenges is the ability to prevent failures by managing corrosion. Most practicing engineers and technologists do not understand how to actively engage in this urgent economic and environmental issue.

This book provides an overview of a comprehensive range of corrosion inhibitors and their applications to combat different forms of known corrosion currently experienced by various industries.

This book is a valuable resource for students, academics, and corrosion engineers alike in combating corrosion.

Table of Contents

Benjamin Valdez Salas, Michael Schorr Wiener, Roumen Zlatev Koytchev, Gustavo López Badilla, Rogelio Ramos Irigoyen, Monica Carrillo Beltrán, Nicola Radnev Nedev, Mario Curiel Alvarez, Navor Rosas Gonzalez, and Jose María Bastidas Rull

CHAPTER

An Overview of Corrosion

Nelson Cheng,[1] Benjamín Valdez Salas,[2] and Patrick Moe[1]

[1]Magna International Pte Ltd., 10H Enterprise Road, Singapore 629834
[2]Universidad Autónoma de Baja California (UABC), Mexicali, Baja California, Mexico

Corrosion is the gradual deterioration of metals caused by the action of air, moisture, or a chemical reaction (such as an acid) on their surface. Rusting of iron, or the forming of brown flaky material on iron objects when exposed to moist air, is the most common example of metal corrosion.

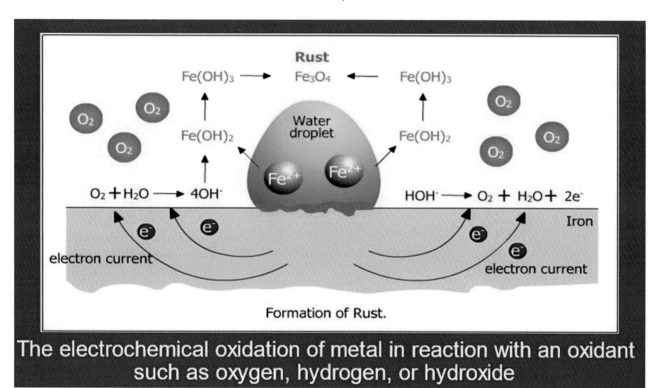

Formation of Rust.

The electrochemical oxidation of metal in reaction with an oxidant such as oxygen, hydrogen, or hydroxide

Corrosion is a natural process that creates a chemically stable oxide from a refined metal. Chemical or electrochemical reactions with the environment slowly destroy materials (usually metals).

In its most common usage, this refers to the electrochemical oxidation of metal in reaction with an oxidant such as oxygen, hydrogen, or hydroxide. A well-known example of electrochemical corrosion is rusting, which forms iron oxides. As a result of this type of damage, oxides or salts of the original metal are produced, resulting in distinctive orange colouration.

In addition to reducing strength, appearance, and permeability to liquids and gases, corrosion reduces the useful properties of materials and structures.

The corrosion process is greatly influenced by exposure to certain substances, but many structural alloys simply corrode from moisture in the air. Surface corrosion can occur uniformly over a wide area, or it can be concentrated locally to form a pit or crack. Surfaces exposed to corrosion are affected by this diffusion-controlled process. Because of this, passivation and chromate conversion can increase the corrosion resistance of a material by reducing the activity of its exposed surface. Some corrosion mechanisms, however, are less visible and less predictable.

Corrosion is an electrochemical phenomenon; its chemistry is complex. During corrosion, an anode occurs at a particular spot on the surface of an iron object. In moist air conditions, electrons from this anodic spot move through the metal to another spot on the metal and reduce oxygen there in presence of H^+ (which is formed when carbon dioxide from the air dissolves into the water because of the dissolution of carbon dioxide from the air into water). In addition to hydrogen ions, other acidic oxides may dissolve into water, releasing hydrogen ions. The said spot acts as a cathode.

1.1. Mechanism of Corrosion

Corrosion, also known as rust, is a chemical or electrochemical degradation of metals that happens because of interactions with the environment. It is an oxidation-reduction process that destroys iron exposed to moisture in the air.

The reaction causes damage to equipment, which results in large costs for many companies to conduct maintenance and repair works.

When ferrous metals are exposed to O_2 and H_2O, a reaction will take place over time, which forms rust in a reaction described as follows:

Iron is first oxidised to iron (II) ions, Fe^{2+}, and oxygen from the air is reduced to hydroxide ions (OH^-). The oxidation-reduction reaction takes place via two separate but simultaneous half-reactions as shown:

Oxidation half-reaction: $Fe\ (s) \rightarrow Fe^{2+}\ (aq) + 2e^-$

Reduction half-reaction: $O_2\ (g) + 2H_2O\ (l) + 4e^- \rightarrow 4OH^-\ (aq)$

Combining the half-reactions from the first step gives a balanced chemical equation for the overall reaction between iron, oxygen, and water:

$2Fe\ (s) + O_2\ (g) + 2H_2O\ (1) \rightarrow 2Fe^{2+}\ (aq) + 4OH^-\ (aq)$

Next, iron (II) hydroxide reacts further with oxygen and water to form hydrated iron (III) oxide ($Fe_2O_3 \cdot n\ H_2O$), which is a flaky reddish-brown solid known as rust:

$4Fe\ (OH)_2\ (s) + O_2\ (g) + XH_2O\ (l) \rightarrow 2Fe_2O_3 \cdot (X+4)\ H_2O\ (s)$ [Rust].

Mechanism of Corrosion

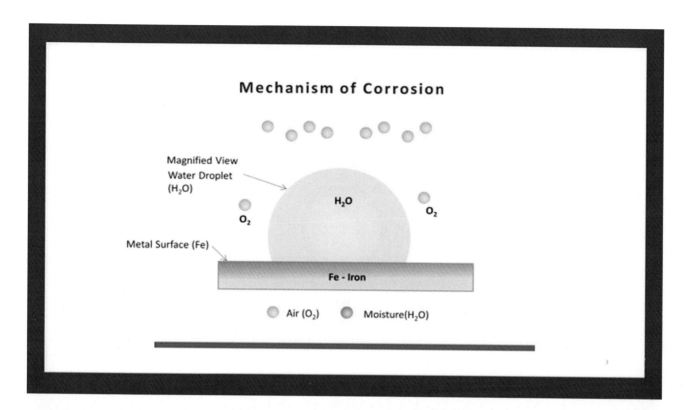

Mechanism of Corrosion

- Fe changes to Fe^{2+} at anode point.
- $Fe \rightarrow Fe^{2+}$ Ferrous ions.

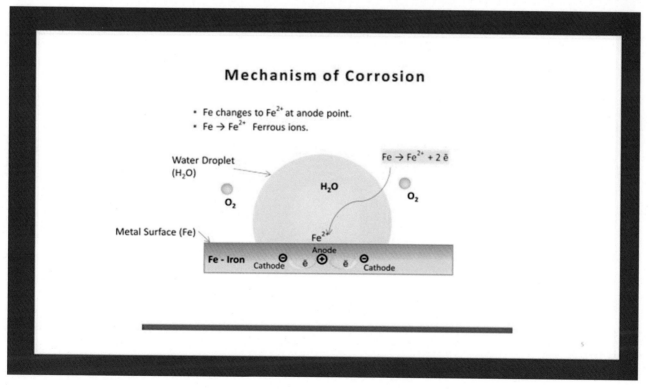

Mechanism of Corrosion

- Cathode accepts ē and O_2 reacts with H_2O to form OH^-
- $2\,\bar{e} + \frac{1}{2}O_2 + H_2O \rightarrow 2OH^-$

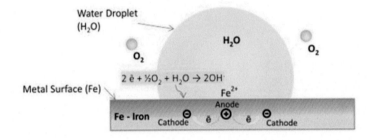

Mechanism of Corrosion

- Fe^{2+} and OH^- react at anode point to form $Fe(OH)_2$
- $Fe^{2+} + 2OH^- \rightarrow Fe(OH)_2$

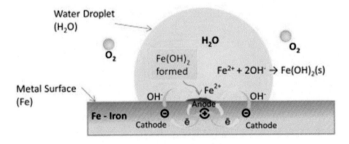

Mechanism of Corrosion

- $Fe(OH)_2$ further react with O_2 and H_2O to form $Fe(OH)_3$ which is rust.

- $4Fe(OH)_2 + O_2 + 2H_2O \rightarrow 4Fe(OH)_3$ Rust

2

An Overview of Types of Corrosion

Nelson Cheng,[1] Benjamín Valdez Salas,[2] and Patrick Moe[1]

[1]Magna International Pte Ltd., 10H Enterprise Road, Singapore 629834
[2]Universidad Autónoma de Baja California (UABC), Mexicali, Baja California, Mexico

This chapter covers the different types of corrosion encountered in generally various types of industries. Corrosion can be classified by the forms in which it manifests itself, based on the appearance of the corroded metal. Observation alone can identify each form. It is usually sufficient to use the naked eye in most cases, but sometimes magnification is helpful. Observing corrosion-damaged test specimens or failed equipment can often provide valuable information for solving corrosion problems. The importance of examination before cleaning cannot be overstated. There are thirteen types of corrosion, and some of them are unique, but all of them are interconnected in some way.

2.1. Atmospheric Corrosion

Atmospheric corrosion occurs when electrolytes interact with metals. As a result of the moisture present in the atmosphere, rainwater, etc., exposed metal surfaces begin to corrode.

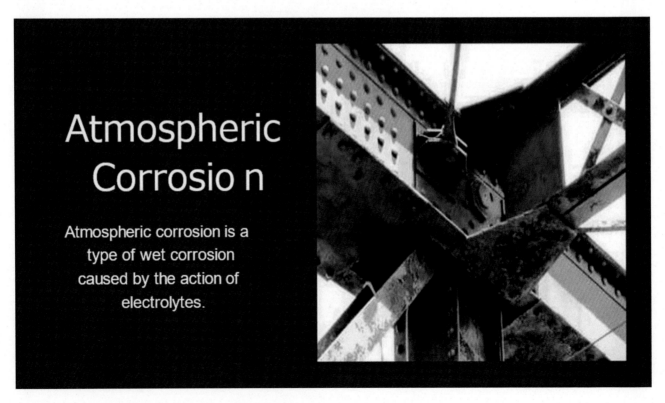

2.1.1 The Mechanism of Atmospheric Corrosion

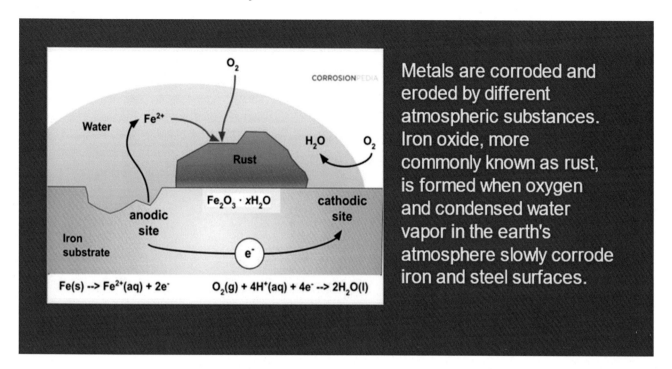

Metals and non-metals are corroded and eroded by different atmospheric substances. Iron oxide, more commonly known as rust, is formed when oxygen and condensed water vapour in the earth's atmosphere slowly corrode iron and steel surfaces. The microstructure of metals is altered by corrosion, reducing their mechanical strength and useful life dramatically.

Atmospheric corrosion is primarily caused by moisture from fog, dew, precipitation, and relative humidity. Corrosion is not caused by oxygen or carbon dioxide in a completely dry atmosphere. The formation of electrolytes in industrial atmospheres can aggravate corrosion caused by sulphur and chlorine salts. There is also an effect of air pressure and ambient temperature on corrosion. High temperatures can cause some electrolytes to become highly reactive. There are factors specific to each metal that determine the critical humidity for corrosion.

There are several types of corrosion damage, but atmospheric corrosion is the most prevalent. The deterioration is widespread, and it affects both indoor and outdoor installations, such as utility lines, industry, vehicles, and residential buildings.

2.2. Galvanic Corrosion

When two dissimilar metals are immersed in a corrosive or conductive solution, there is a potential difference. This potential difference produces electron flow between these metals when they are placed in contact (or otherwise electrically connected). Usually, the corrosion rate of a less corrosion-resistant metal increases, while the attack rate of a more corrosion-resistant metal decreases when these metals are in contact. A metal that is less resistant becomes anodic, while a metal that is more resistant becomes cathodic. Usually, the cathode or cathodic metal corrodes very little or not at all in this type of couple. Galvanic corrosion is two-metal corrosion caused by electric currents and dissimilar metals. The corrosion is electrochemical; though, for clarity, we will refer to it as galvanic corrosion.

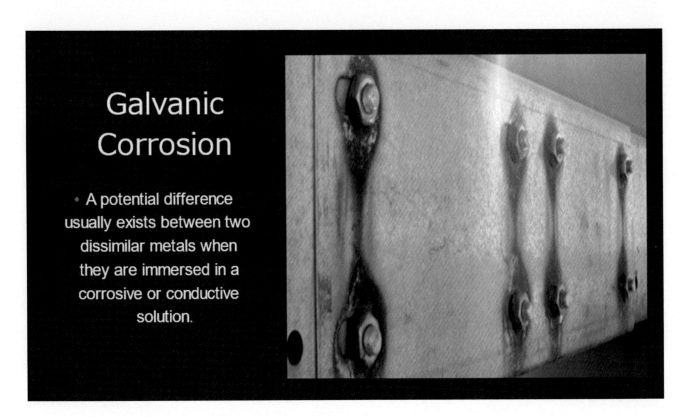

Galvanic Corrosion

- A potential difference usually exists between two dissimilar metals when they are immersed in a corrosive or conductive solution.

2.2.1 The Mechanism of Galvanic Corrosion

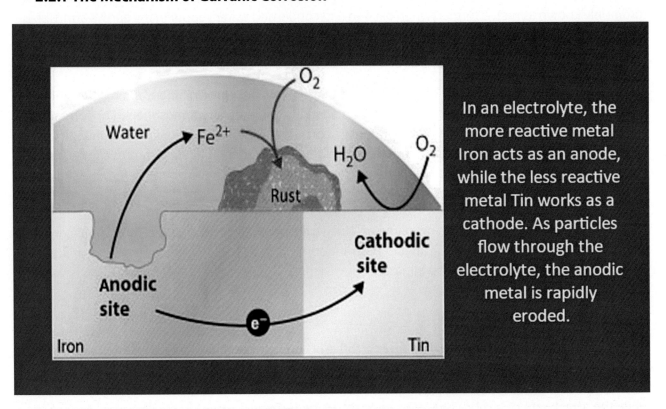

In an electrolyte, the more reactive metal Iron acts as an anode, while the less reactive metal Tin works as a cathode. As particles flow through the electrolyte, the anodic metal is rapidly eroded.

It is a galvanic cell's potential difference between its two metals that make it work. This potential difference causes electrons to flow within the cell. An electrode's oxidation potential refers to its ability to lose electrons and become oxidised. Metals with a higher oxidation potential give up their electrons more easily.

2.3. Crevice Corrosion

On metal surfaces exposed to corrosives, intense localised corrosion often occurs in crevices and other shielded areas. There are several causes of this type of attack, including holes, gasket surfaces, lap joints, surface deposits, and crevices under bolt and rivet heads. Therefore, this type of corrosion is called crevice corrosion or deposit corrosion.

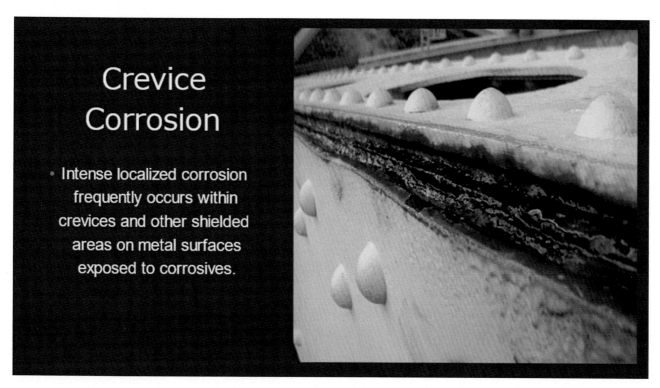

Crevice mechanisms are extremely complex. The mechanism relies heavily on concentration cells. A corrosion cell is formed when metal ions differ from those outside the crevice. The cathode is the area with low metal concentration; the anode is the area with high metal concentration.

When oxygen concentrations outside and inside a crevice differ, a differential cell is formed, resulting in crevice corrosion.

In crevice corrosion, stagnant solutions attack metal surfaces in crevices, such as around nuts and rivet heads. As dust, sand, and other corrosive substances accumulate on surfaces, water accumulates and corrodes the part. This can happen between metals or between metals and non-metals. Chemical concentration gradients cause damage to metallic parts.

Outside the crevice, oxygen causes an electrochemical concentration cell. This is a differential aeration cell that uses oxygen as the air source. There is an increase in pH and oxygen content in crevices (the cathode).

The corrosion is worsened by chlorides' higher electrochemical concentration on the inside. When ferrous metals are present, ferrous ions react with chlorides to form ferric chloride, which attacks stainless steel. In this way, both the oxygen concentration and the pH concentration remain lower in the crevice than in the solution formed by the water on the metal. As with pitting corrosion, the propagation mechanism is similar. The mechanism is as follows:

Anode $\qquad$ $M \rightarrow M^{++} + 2e^-$

Cathode $\qquad$ $\frac{1}{2} O_2 + H_2O + 2e^- \rightarrow 2OH^-$

Oxygen reduction outside the crevice.

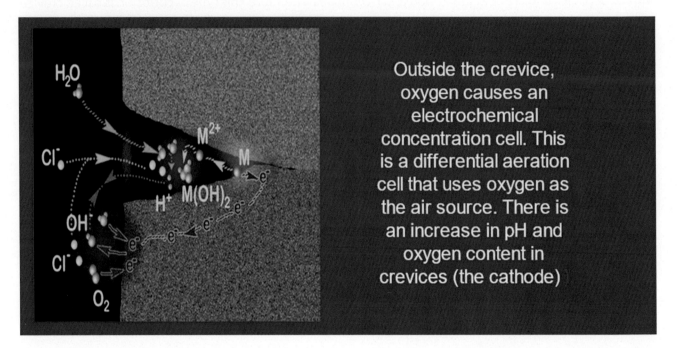

2.4. Erosion Corrosion

Because of the relative movement between metal surfaces and corrosive fluids, erosion corrosion occurs. By abrasion of fast-moving fluids, the surface of metal gradually deteriorates, and cavities form. Metal tubes that contain moving fluids are commonly affected by this type of corrosion.

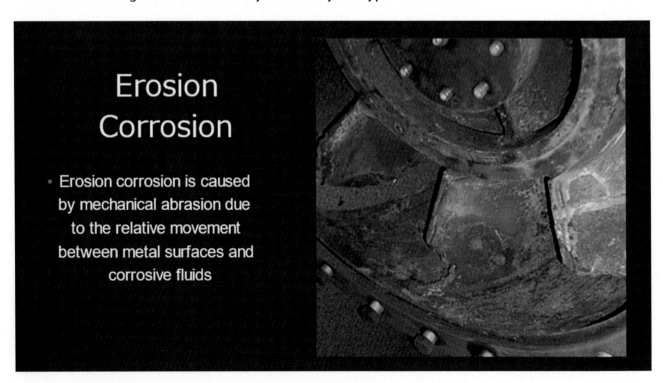

Fluid flows through a pipe at high velocity, causing friction and physical effects on the surface, along with corrosive action from the fluid. Consequently, metal is lost more quickly.

A protective film usually covers the metal, which is the first part to erode. As soon as the film disappears, the bare metal becomes vulnerable to corrosion. Constriction areas are prone to this type of corrosion. A high rate of flow can be observed in places where there are blockages, inlet ends, pump impellers, etc.

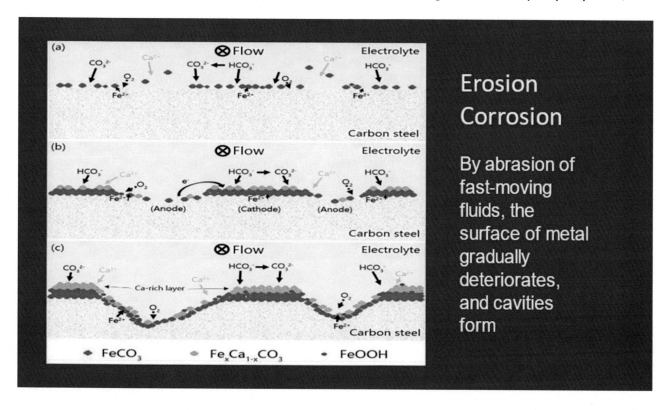

A metal surface can be corroded or eroded because of erosion and erosion combined, and erosion-corrosion occurs when a turbulent fluid flows rapidly over it. Turbulence is caused by pitting, which is often found on the inner surface of pipes.

During turbulent conditions, erosion rates increase, which can lead to leaks in tubes and pipes.

Corrosion caused by erosion can also be caused by poor workmanship. Without removing inner burrs during installation, these inner burrs cause localised turbulence and impede fluid flow.

2.5. Selective Corrosion

Selective corrosion occurs in alloys when one of the component metals is de-alloyed by the corrosion environment. The zinc in brass alloy pipes is commonly de-alloyed, resulting in this type of corrosion. Nickel is also de-alloyed by selective corrosion in copper-nickel alloy tubes.

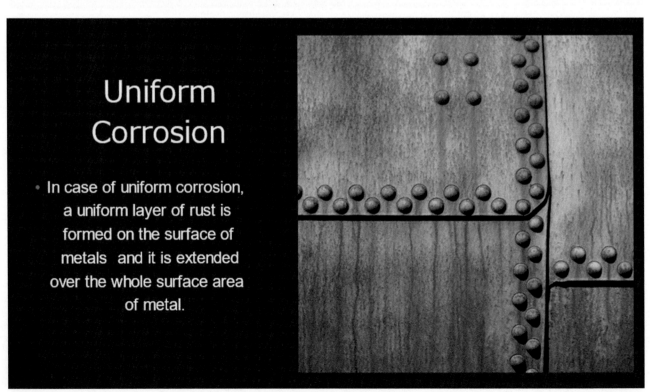

2.6. Uniform Corrosion

The uniform corrosion of metal results in a uniform layer of rust covering the entire surface area of the metal. Metals without a surface coating are susceptible to this type of corrosion. Metals such as aluminium, zinc, lead, etc. are commonly affected by uniform corrosion.

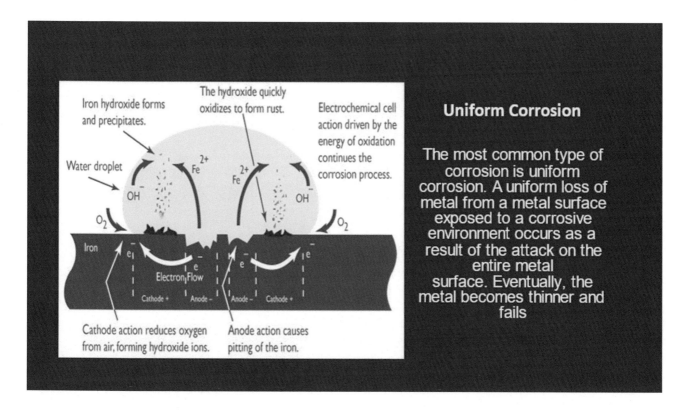

Uniform Corrosion

The most common type of corrosion is uniform corrosion. A uniform loss of metal from a metal surface exposed to a corrosive environment occurs as a result of the attack on the entire metal surface. Eventually, the metal becomes thinner and fails

2.7. Pitting Corrosion

Pitting is the formation of rust pits or holes on the surface. Pitting corrosion is a localised form of corrosion where the corrosion is limited to small areas. The shapes of rust pits may not be similar, but in most cases, they are hemispherical in shape.

Pitting corrosion occurs when the protective oxide layer of the surface gets damaged or because of structural defects in metal. It is considered more dangerous because it causes the failure of structure with a relatively low overall loss of material. It can be observed in steel, aluminium, nickel alloys, etc.

2.8. Fretting Corrosion

Fretting corrosion occurs at the contact area of the two materials that are joined together. This is developed when the contact area is subjected to slips and vibrations. This type of corrosion can be seen in bolted and riveted joints, clamped surfaces, etc.

2.8.1 Mechanism of Fretting Corrosion

The wear caused by fretting normally occurs between surfaces that shouldn't be moving against one another. Vibration or elastic deformation of the involved body is often responsible for this microscopic relative movement.

Contact surface asperities adhere to each other, which is then broken by the small movement. During this process, wear debris is formed. The process of fretting corrosion occurs when debris and/or surface undergo chemical reactions, mainly oxidation.

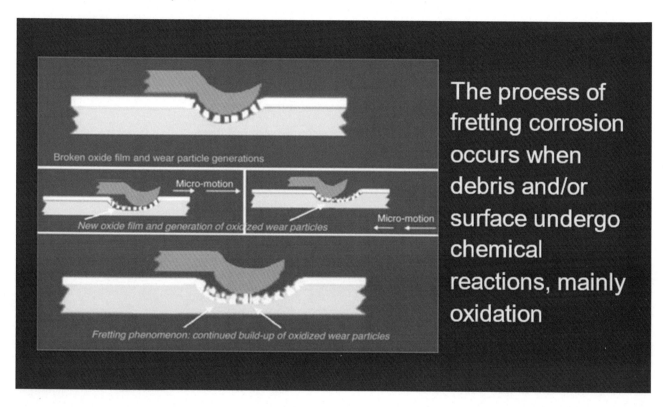

2.9. Stress Corrosion

Stress corrosion is caused by the combined action of a corrosive environment and mechanical stress on the surface of the material. In its initial stage, small cracks are developed, and these finally lead to the failure of the whole structure. This type of corrosion can be seen in stainless steel when they are stressed in chloride environments, in brass materials when they are stressed in the presence of ammonia, etc.

Corrosion in this form can occur either as intergranular stress corrosion cracking (IGSCC) or as transgranular stress corrosion cracking (TGSCC).

Intergranular stress corrosion cracking (IGSCC) occurs when cracks form along the grain boundaries. A fracture (crack) forms through the grains of material (and not along its boundaries) in the case of transgranular stress corrosion cracking (TGSCC).

2.9.1 The Mechanism of Stress Corrosion

In stress corrosion, a relatively inert material corrodes because of applied stress. There are two types of stress: externally applied and residual. Corrosion of this type may not occur until external stress is applied, which makes it particularly dangerous.

Stress corrosion cracking is caused by three main factors working together: material, environment, and tensile stress.

The type of material used can cause stress corrosion cracking. Stress corrosion cracking occurs differently in different materials because of their different susceptibilities. As a result of the corrosive environment in which it is operating, poor material selection can lead to stress corrosion cracking.

Chemical species in the service environment can cause stress corrosion cracking in different materials. As a result, material and environment selection should be considered together to avoid stress corrosion cracking.

The material experiences stress or strain because of residual stress or direct application of stress. Stress corrosion cracks propagate primarily because of static stress. Stress corrosion cracks must also be characterised by factors related to materials and environment to be considered stress corrosion cracks.

2.10. Intergranular Corrosion

Intergranular corrosion is the corrosion that occurs along the grain boundaries and the grains are not affected in this case. It is caused when there is a noticeable difference in reactivity against impurities that exist between grain boundaries and grains. This difference in reactivity occurs because of defective welding, heat treatment of stainless steel, copper, etc.

2.11 Corrosion Fatigue

Fatigue of materials is defined as the failure of a material because of repeated application of stress. When the fatigue of metal is developed in a corrosive environment, then it is called corrosion fatigue. Corrosion fatigue is caused by cracks developing simultaneously with corrosion and cyclic stress. It is usually associated with rapidly fluctuating stresses that are well below tensile strength. As stress increases, the number of cycles required to cause fracture decreases.

2.11.1 Mechanism behind Corrosion Fatigue

Corrosion fatigue occurs in corrosive environments. During corrosion and cyclic loading, a material is mechanically degraded. It occurs when alternating stresses and corrosive environments are combined. Fatigue is believed to cause a rupture of the passive film, which accelerates corrosion.

An alternating tensile/pressure load increases corrosion fatigue crack growth rate. In contrast with ordinary fatigue, corrosion fatigue depends heavily on the frequency of the alternating load. Corrosion has a greater effect on crack growth rates at low frequencies because of the prolonged contact between the crack surface and the electrolyte. Corrosion fatigue cracks grow slower with higher frequency.

In their lifetime, most engineering structures experience some form of alternating stress and are exposed to harmful environments.

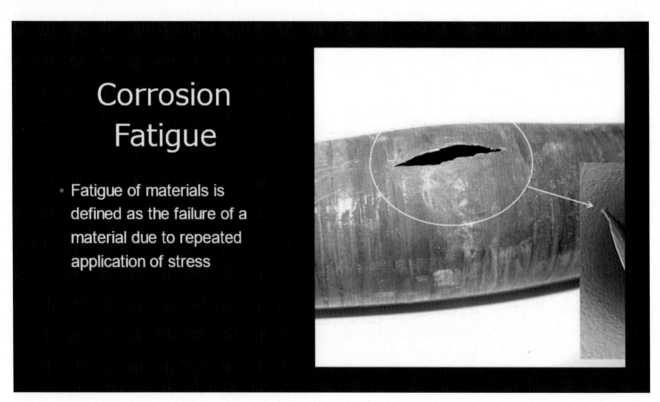

An Overview of Corrosion Inhibitors

Nelson Cheng,[1] Benjamín Valdez Salas,[2] and Patrick Moe[1]

[1]Magna International Pte Ltd., 10H Enterprise Road, Singapore 629834
[2]Universidad Autónoma de Baja California (UABC), Mexicali, Baja California, Mexico

Definition of Corrosion Inhibitors

Corrosion control of metallic infrastructure can be achieved by the implementation of different technologies including preventive actions: the correct selection of metals or alloys, surface engineering for protective coatings such as organic paints, cladding, inorganic chemical conversion coatings, etc., cathodic protection, both sacrificial anode and impressed current, anodic protection, and corrosion inhibitors.

Corrosion inhibitors (CIs) are chemical substances that can be added or incorporated to the corrosive media in small quantities but enough to provide the metal with an efficient and safe corrosion protection. Other important characteristics to fulfil for are they do not modify the media physic properties of the media in which are incorporated, and they are non-toxic and at a reasonable cost.

3.1 How It Works

The CI interacts with the electrochemical or chemical reactions occurring in the metal corrosion process, diminishing the corrosion rate to acceptable levels. From an electrochemical point of view, CIs can interfere with the cathodic reaction (cathodic inhibitor), anodic reaction (anodic inhibitor), or both (mixed inhibitor). Because the CI must be incorporated in the corrosive media, their application is restricted to semi-closed or closed systems, and its dosage can be in single dose or continuous doses from liquid solution or solid soluble tablets for extended release in aqueous solutions. Vapour corrosion inhibitors (VCI) are chemical substances with vapour pressure values that permit them to be dispersed in closed system where the metallic surfaces are not immersed in an aqueous corrosive environment, but there is a presence of humidity. Examples of these corrosion systems are electrical cabinets, electro-electronic devices, oil containers in ships and storage terminals, vehicles, armament, etc.

3.2 Classification of Corrosion Inhibitor (CIs)

CIs can be of organic or inorganic origin or a mixture of them. They are widely used for protection of cooling or heating water systems, in closed circuit, chemical pickling of metals in acids to avoid metal hydrogen embrittlement, as well as prevent metal loss, washing solutions for degreasing, cooling solutions in the system of pumps, lubricating solutions in polishing systems and metal finishing, among others.

The type of inhibitor is selected based on the metal or system of metals, alloys, contacts of different metals, etc. Likewise, the corrosive solution to be treated with the inhibitor and its compatibility with this electrolyte are considered. The CIs protect the metal as they are adsorbed on their surface and thus block the attack of the corrosive environment through the film formed on the metal.

3.3 Corrosion Inhibitor Inhibition Efficiency

Efficiency of the inhibitor is characterised by the value of the degree of protection (P), as well as by the protective action coefficient γ—equation (1)—and can be calculated using equation (2):

$$\gamma = C_o/C_i, \qquad (1)$$

$$P = [(C_o - C_i)/C_o] \times 100 \ (\%), \qquad (2)$$

where C_o and C_i are the corrosion rates of the metal in the media with and without corrosion inhibitor respectively.

3.4 Measurement of Corrosion Rate

Corrosion rate can be expressed in different units depending on the technical system and type of corrosion (uniform, localised, etc.). In this way, it can be understood as the increase in the depth of corrosion penetration per unit of time (μm/year) or by the loss of mass of the material per unit area and unit of time (g/m^2.year), etc. The effect of corrosion can vary with time, and its development can be different at all points of the corroded metal. Therefore, reports regarding the rate of corrosion must be accompanied by information on the type of corrosion effect, its change over time, and its location.

3.5 How Anodic Corrosion Works

Anodic corrosion inhibitors reduce the partial anodic reaction of the corrosion process by passivating the metal or forming a film of slightly soluble products on the anodic sites of the metal surface. There are two groups of inhibitors: (a) passivators, oxidants—for example, oxygen (O_2), or other strong oxidants, such as hydrogen peroxide (H_2O_2), chromates (CrO_4^{-2}), dichromate ($Cr_2O_7^{-2}$), nitrates (NO_3^-), nitrites (NO_2^-), molybdates (MoO_4^{-2}), etc.; and (b) protective film-forming inhibitors—for example, phosphates (PO_4^{-3}), carbonates (CO_3^{-2}), benzoates ($C_6H_5COO^-$), etc.

3.6 How Cathodic Corrosion Works

Cathodic inhibitors decrease the cathodic partial reaction of the corrosion process, affecting its effectiveness or blocking the cathodic sites on the metal surface. They are reducing chemical compounds (Na_2SO_3, N_2H_4, and others), which can bind oxygen and reduce its concentration in the electrolyte solution, when the corrosion process is carried out with depolarisation by oxygen. Another type of inhibitor in this group is made up of compounds ($AsCl_3$, $Bi_2[SO_4]_3$) that hinder the hydrogen evolution process.

3.7 How Mixed Corrosion Works

Mixed inhibitors retard both the anodic and cathodic reactions. They are chemical compounds such as polyphosphates, silicates, and various organic substances that are adsorbed on the metal surface to hinder the development of both electrochemical reactions.

The various types of inhibitors are widely used in the protection against corrosion of metallic structures in different aggressive environments: acids and neutral aqueous solutions of salts, corrosive atmospheres of industrial, urban or marine type, etc.

In corrosive acid environments, many organic compounds are used, such as amines, mercaptans, and other types of sulphur compounds (thiols, R-SH), nitrogenous heterocyclic compounds, alcohols, acetylene derivatives (HCCH, C_2H_2), aldehydes (RCH=O), which generally have mixed inhibitory behaviour. The degree of protection of organic inhibitors depends on their nature, geometric structure of the molecule, chemical composition, character and position of the functional groups, type of bond, as well as the concentration of the inhibitor in the acid solution, temperature, and type of acid involved in the corrosive process. Inhibitors for acid media are widely used in chemical and electrochemical pickling processes, acid cleaning of heat exchanger equipment, tanks and containers used to transport acids, protection of equipment in contact with acid industrial processes, among others.

Inhibitors for water and neutral aqueous solutions of salts, which are used in industries as a means of anticorrosive protection, are of the oxidising type—passivators, chromates, nitrates, molybdates, vanadate, and film formers such as phosphates, polyphosphates, silicates, etc., amines, and others. It is worth mentioning that anodic corrosion inhibitors are very dangerous when their concentration is not sufficient to ensure anticorrosive protection of the metal, so they can cause severe localised corrosion such as ulcers and pitting. The critical concentration of most of these ICs ensures a high degree of protection. It depends on the temperature and pH of the medium, especially when it contains aggressive ions such as chlorides or sulphates.

Inhibitors for protection against atmospheric corrosion can be non-volatile, contact, and volatile. Contact inhibitors (for example, sodium nitrite, $NaNO_2$ for carbon steels and grey steels) are applied on the metal surface or on the packaging paper for preservation. These ICs form a passive layer on the metal surface with good protective properties; however, their action is limited only at the place of contact with the metal itself. Contact corrosion inhibitors are used for temporary protection against corrosion—for example, during storage, transportation, the conservation of equipment or works of art.

3.8 How VCI Works

Volatile corrosion inhibitors (VCI), vapour phase corrosion inhibitors, nitrites, carbonates of secondary amines, salts of carboxylic acids, and others form and exert a very high vapour pressure, rapidly saturating the atmosphere around the metal and subsequently dissolving, absorbed on its surface, ensuring a protective environment against atmospheric corrosion.

The use of volatile inhibitors requires very good cleaning of the metal surface to be protected, as well as hermetic packaging. These inhibitors are very modern and effective means of anticorrosive protection and are being used to protect metal parts or equipment that include metal parts or components during storage, transportation, or conservation for a certain time. They are required to act on the metal in a closed space (for example, inside a container, closed boxes, polyethylene, or paper bags), since the inhibitor evaporates in the surrounding atmosphere, and for it to be sufficient and efficient its concentration, it is required to limit the space. Being adsorbed on the metal, the volatile inhibitor blocks the metal surface and stops the attack of the corrosive environment.

Regardless of the wide possibilities for the use of inhibitors in practice, it is necessary to remember that their action is specific in relation to the diversity of metals and alloys that exist; likewise, their introduction into the corrosion system should not have an unfavourable influence on the technology, human health, and the natural environment.

That is why the large manufacturers of these inhibitors such as Magna International use green chemistry processes and produce VCIs that are friendly to the environment and human health and have ruled out those compounds that are harmful from their formulations. The use of VCI is economically feasible in a constant volume corrosion system or with little change in the corrosive solution.

3.9 References

Raicho Raichev, Lucien Veleva y Benjamín Valdez, Corrosión de metales y degradación de materiales. Principios y prácticas de laboratorio. Editorial UABC, ISBN 978-607-7753-07-0, 2009. 380 pp.

CHAPTER

An Overview of Types of Corrosion Inhibitors

Nelson Cheng,[1] Benjamín Valdez Salas,[2] and Patrick Moe[1]

[1]Magna International Pte Ltd., 10H Enterprise Road, Singapore 629834
[2]Universidad Autónoma de Baja California (UABC), Mexicali, Baja California, Mexico

4.1 Types of Corrosion Inhibitors

Based on how they function, the environment they are exposed to, and how they operate, corrosion inhibitors can be categorised into four categories—namely, anodic inhibitors, cathodic inhibitors, mixed inhibitors, and volatile corrosion inhibitors (VCIs).

4.1.1 Anodic Inhibitor

On the surface of the metal, these corrosion inhibitors form a thin protective oxide layer. During this reaction, the metallic surface undergoes a big anodic shift, turning into a passivation zone. The passivation area reduces corrosion on the metal. Chromate, nitrite, orthophosphate, and molybdate are few examples of anodic inhibitors.

4.1.2 Cathodic Inhibitor

A cathodic inhibitor slows the cathodic reaction or selectively precipitates on the cathodic regions of the metal to prevent diffusion of eroded elements to the metal surface. Cathodic inhibitors include sulphite and bisulphide ions, which can react with oxygen to form sulphates. The catalysed redox reaction that is catalysed by nickel is another example of a cathodic inhibitor.

4.1.3 Mixed Inhibitors

A corrosion inhibitor film can also be formed on the metal surface when these types of inhibitors are used. As well as reducing anionic reactions, they also reduce cationic reactions. During this process, a precipitate is formed on the surface of the metal to achieve this process. By mixing inhibitors with water, metals are prevented from rusting when they meet silicates and phosphates.

4.1.4 Volatile Corrosion Inhibitor

Condenser tubes in boilers can be protected from corrosion by volatile corrosion inhibitors. They are also referred to as vapour phase inhibitors, or VPIs.

To prevent corrosion, VCIs change the pH of the exterior atmosphere to a less acidic state. A few examples are morpholine and hydrazine, which are used to prevent the corrosion of condenser pipes.

4.2 Corrosion Inhibitors and Their Uses

The use of corrosion inhibitors in commercial, industrial, and process environments is wide ranging. Below are a few of these uses.

- Corrosion inhibitors are used to prevent rusting and anodic corrosion of metals. Generally, this is achieved by coating the metal surface with chromate.
- Cathodic corrosion can be prevented by using oxygen scavengers as CIs to react with dissolved oxygen in the environment.
- Keeping fuel pipelines from corroding and rusting is very important. Thus, these pipelines are more secure, and accident risks are reduced with CIs.
- Corrosion can occur in metal pipes in heating systems. It is also important to secure these pipes with CIs.

4.2.1 A corrosion inhibitor can be classified according to its protection mode as follows:

- passivators containing chemical agents (nitrites, chromates, zinc molybdate, etc.)
- adsorption inhibitors (quinolines, sulphur atoms in metal compounds, nitrogen atoms in amines, oxygen atoms in aldehydes)
- inhibitor of film formation (zinc and calcium salts, benzoate, etc.)
- vapour phase corrosion inhibitors (dicyclohexylamine chromate, benzotriazole, tolyltriazole, phenylthiourea, cyclohexylamine, dicyclohexylamine nitrite, etc.).
- corrosion inhibitors that are volatile (molybdenum oxide, salts of dicyclohexylamine, cyclohexylamine, and hexamethylene amine)
- synergistic inhibitors (chromate-phosphates, polyphosphate-silicate, zinc-tannins, and zinc-phosphates)
- precipitation inhibitors (sodium silicate)
- green corrosion inhibitors (amino acids, alkaloids, pigments, and tannins)
- organic inhibitors (amines, aldehydes, alkaloids, nitroso compounds)
- inorganic inhibitors (As_2O_3 and Sb_2O_3)
- alkaline inhibitors (thiourea, substituted phenols, naphthol, β-diketone, etc.)
- inhibitors with neutral properties

4.3 Inhibitor Carriers

The terms 'oil soluble', 'water soluble', 'oil-soluble, water dispersible', etc. are commonly used to describe corrosion inhibitors. When such terms are used, they are generalisations, not rigorous descriptions. The only difference between a water-soluble inhibitor and an oil soluble inhibitor is that an oil-soluble inhibitor may partition more towards the oil phase.

A single compound has a clearly defined partitioning between two phases. Although many commercial corrosion inhibitors are complex mixtures of many compounds, each with its own partition coefficient, it is important to remember that many are complex mixtures of many compounds. The partitioning coefficient of commercial corrosion inhibitors differs for each component, not for the entire product. Generally, organic inhibitors are more soluble in aromatic hydrocarbons than aliphatic hydrocarbons, and in long chain aliphatic hydrocarbons than short chain aliphatic hydrocarbons. Consequently, the partition coefficients of each oil of interest must be measured.

4.4 Selection of Corrosion Inhibitors Based on an Evaluation Process

Screening out unsuitable inhibitors should be done by using the simplest tests first. During initial screening tests, inhibitors should not be carried forward if they perform poorly. Inhibitors that don't perform well in early screening tests might do well in the real system, but users rarely have the resources to test all possible inhibitors. Even if some good inhibitors are excluded, the inhibitor user must use test procedures that rigorously exclude inferior inhibitors.

4.5 An Overview of the Inhibitor Selection Process

The choice of physical properties is the first step in selecting an inhibitor. Does the inhibitor have to be solid or liquid? What is the importance of melting and freezing points? Is temperature and time degradation critical? Does it need to be compatible with other additives in the system? Are there any specific solubility requirements? It is important to define the domain of possible inhibitors from this extensive list. Any new system must begin with this step. Physical measurements like these are routinely performed as part of minimal quality acceptance testing.

Inhibitor evaluation involves designing experiments that simulate the conditions of the real world. In addition to metal properties and corrosive environment chemistry, temperature, pressure, and velocity must be considered. Usually, corrosion failures are localised and caused by micro conditions at the site of failure. The most severe conditions that can occur in the system must be included in adequate testing, not only macro or average conditions. Heat exchangers with hot spots and weld beads with highly turbulent flow are examples of microenvironments.

4.6 Materials Used in the Testing Process

4.6.1 Performing Metal Testing

Test specimens should be made from the same metal as the one to be protected; even small differences in metal chemistry can affect inhibitor performance significantly. Since inhibitor performance varies greatly on different metals, ranking inhibitors based only on one metal is not universal. The differences between 'same' metals are less obvious. Surface condition, grain size, and orientation are examples of non-chemical differences. Attempts should be made to prepare a surface that is comparable to that of the modelled system. As a rule, minimal cleaning includes a solvent wash to degrease the sample, except in special tests. The results of tests can be markedly affected by more vigorous cleaning procedures such as bead blasting or acid activation, despite improved reproducibility. Electrochemical measurements often require activating test specimens in acid. It is necessary to remove any passive or protective oxide layer to reach metal solution equilibrium as quickly as possible. Preparation, cleaning, and evaluation of corrosion test specimens are described in ASTM G1.

4.6.1 Methods for Measuring Metal Losses

Metal loss can be measured gravimetrically, volumetrically, or radiometrically. The most common method for testing inhibitors is gravimetric or weight loss. The use of volumetric methods is associated with inspection or monitoring techniques such as ultrasonic inspection and electric resistance (ER) probe monitoring, although both are sometimes used in the evaluation of long-term inhibitors. Methods such as radiometric measurements are used to monitor thin layer activation but could also be used for the assessment of inhibitors. A weight loss test used to evaluate oilfield inhibitors is the corrosion wheel test. It is important to visually inspect coupons from weight loss experiments for pitting or edge damage caused by localised corrosion. There can be a simple analysis such as 'none, some, or lots', or there can be a more detailed analysis like counting and depth measurement. Pitting corrosion can be evaluated using ASTM G46 (Practice for Examination and Evaluation of Pitting Corrosion).

4.6.1 Methods Based on Electrochemistry

There are two major benefits to electrochemical testing, one major limitation, and one lesser limitation. Mechanistic information and short measurement times are the benefits. A conductive, corrosive environment is a severe limitation. The requirement for a corrosion model is less burdensome from a testing perspective. Inhibitor performance can be characterised rapidly using these techniques. In contrast to weight-loss methods that can take days to determine, corrosion rates can be determined electrochemically in minutes. Changes in inhibitor performance over time are readily measurable using electrochemical methods because of their near-instantaneous nature. Experimental questions related to inhibitor persistence and incubation time are thus more accessible, and experiments relating to velocity effects become less time-consuming.

4.7 Inhibitor Selection for Corrosion Inhibition

There are only a few rules, equations, or theories to guide the development or use of inhibitors. A synergy is often present between different inhibitors and the environment being controlled, and mixtures are common in commercial formulations. When selecting an inhibitor system, it is important to consider how it is usually implemented.

Listed below are the names of inhibitors, their inhibition efficacy on the desired metals, and their concentrations in different media.

Table 1. The use of inhibitors to protect some corrosive systems

System	Inhibitor	Metals	Concentration
Acids			
HCl	ethylaniline	Fe	0.5%
	mercaptobenzotriazole	..	1%
	pyridine + phenylhydrazine	..	0.5% + 0.5%
	rosin amine + ethylene oxide	..	0.2%
Sulphuric	phenylacridine	..	0.5%

System	Inhibitor	Metals	Concentration
Phosphoric	sodium iodide	..	200 ppm
Others	thiourea	..	1%
	sulphonated castor oil	..	0.5–1.0%
	arsenic oxide	..	0.5%
	sodium arsenate	..	0.5%
Water			
Potable	calcium bicarbonate	steel, cast iron	10 ppm
	polyphosphate	Fe, Zn, Cu, Al	5–10 ppm
	calcium hydroxide	Fe, Zn, Cu	10 ppm
	sodium silicate	..	10–20 ppm
Cooling	calcium bicarbonate	steel, cast iron	10 ppm
	sodium chromate	Fe, Zn, Cu	0.1%
	sodium nitrite	Fe	0.05%
	sodium phosphate monobasic	..	1%
	morpholine	..	0.2%
Boilers	sodium phosphate monobasic	Fe, Zn, Cu	10 ppm
	polyphosphate	..	10 ppm
	morpholine	Fe	variable
	hydrazine	..	O2 scavenger
	ammonia	..	neutraliser
	octadecylamine	..	variable
Engine coolants	sodium chromate	Fe, Pb, Cu, Zn	0.1–1%
	sodium nitrite	Fe	0.1–1%
	borax	..	1%
Glycol/water	borax + MBT	All	1% + 0.1%
Oilfield brines	sodium silicate	Fe	0.01%
	quaternaries	..	10–25 ppm
	imidazoline	..	10–25ppm
Seawater	sodium silicate	Zn	10 ppm
	sodium nitrite	Fe	0.5%
	calcium bicarbonate	All	pH dependent
	sodium phosphate monobasic + sodium nitrite	Fe	10 ppm + 0.5%

CHAPTER 5

Overview of Metal Corrosion Rate Calculation

Nelson Cheng,[1] Benjamín Valdez Salas,[2] and Patrick Moe[1]

[1]Magna International Pte Ltd., 10H Enterprise Road, Singapore 629834
[2]Universidad Autónoma de Baja California (UABC), Mexicali, Baja California, Mexico

Most metals undergo a chemical change when they meet certain substances in the air, soil, or water. Corrosion is the result of this process. There are several materials that can cause corrosion, including gases such us oxygen, hydrogen sulphide, NO_x, SO_x, CO_x, polluted, brackish, and saline waters, anions like sulphates and chlorides, among others.

Corrosion of metal results in it being unable to hold the same loads it did before it corroded. It is possible for corrosion to lead to dangerous conditions at a certain point. Bridges, railroad tracks, chemical plants, vehicles, military equipment and armament, energy generation and distribution installations, communication infrastructure and buildings all use metals that can corrode. Because of corrosion, the infrastructure mechanical, aesthetics, and safety properties are diminished with the consequent risks for their operation. Thus, corrosion must be monitored and managed to prevent structural collapse.

5.1 Definition of Rate of Corrosion

The rate of corrosion is the rate velocity at which a metal deteriorates by its interaction with a particular environment. Depending on the type and condition of the metal, as well as the environment, the rate will differ. Corrosion rate is typically calculated using mpy (millimetre per year) as a measurement of the uniform corrosion penetration on the metal surface.

Corrosion rates in the US are normally calculated using mpy (mils per year). Thus, corrosion rate is determined by the number of millimetres (thousandths of an inch) it penetrates each year.

The following information must be collected to calculate corrosion rates:
- loss of weight (decrease in metal weight over the reference period)
- metal density (density of the metal)
- surface area (total initial surface area of the metal piece)
- the reference period (the length of the reference time)

5.2 Corrosion Rate Conversion

Using the following equation, you can convert mils per year to millimetres per year (mm/y):

1 mpy = 0.0254 mm/y = 25.4 micron/y.

To calculate the corrosion rate from metal loss:

mm/y = 87.6 x (W / DAT),

where

W = weight loss in milligrams,

D = metal density in g /cm^3,

A = area of sample in cm^2,

T = time of exposure of the metal sample in hours.

5.3 The Importance of Corrosion Rates

A metal-based structure's lifespan is determined by its corrosion rate. As a result, different metals are used for different purposes and in different environments. A metal structure in a wet environment may require more frequent maintenance than a similar structure in a dry environment. The types of calculations described above are used to develop maintenance schedules.

During the design, maintenance, and until the equipment is scrapped, corrosion is one of the parameters engineers must deal with. Our standard corrosion allowance for carbon steel is 3 mm. In some cases, however, the corrosion allowance should exceed 3 mm because the handling fluid has a high corrosion rate.

5.4 Inhibitor Efficiency

A corrosion inhibitor is a chemical substance that, when added to an environment at a small concentration, effectively reduces corrosion. A measure of this improvement is used to measure the effectiveness of that inhibitor:

IE-Inhibitor Efficiency (%) = 100 X ($CR_{uninhibited}$ − $CR_{1inhibited}$) / $CR_{uninhibited}$,

where

$CR_{uninhibited}$ = corrosion rate of the uninhibited system,

$CR_{1inhibited}$ = corrosion rate of the inhibited system,

IE = (CR-CR_1) / CR.

CI concentration is one of the most important parameters during its application. For a film forming CI, as inhibitor concentration increases, its efficiency increases. For example, a typical good inhibitor might provide 95 per cent inhibition at 0.006 per cent, and 90 per cent at 0.004 per cent. On the other hand, CI having chemical interactions with the metallic surface are used in function of critical concentrations; it means there is a maximum concentration in which CI can protect safely the metal. If critical concentration is exceeded, corrosion can be promoted instead to be inhibited.

To improve inhibitor capabilities, sophisticated corrosion inhibitor test methods have been employed over the years. They typically reproduce the most extreme conditions in a system. Despite their success in elaborate laboratory apparatus, many corrosion inhibitors have not reached comparable performance in the field despite their performance in elaborate laboratory apparatus. Today, it remains difficult to transfer inhibitor performance from the laboratory to the field. When key inhibitor chemistry and corrosion theory factors are considered, it may be possible to correlate laboratory and field performance.

Corrosion rate refers to the uniform destruction of material by corrosion, without pitting or other forms of local attack. Experimentally determined corrosion rates can be used to predict the life of a material in service if the corrosion is uniform.

5.5 Calculation of Corrosion Rate

It is usually expressed as a penetration rate in inches per year or mils per year (MPY) (where mils is 10^{-3} inches). Corrosion rate is calculated by exposing a specimen to corrosive conditions for a predetermined period:

$$ipy = \frac{12W}{TAR}$$

where

W = mass loss in time (T), lb;

T= time, years;

A = Surface area, ft^2;

R = density of material, lb/ft^3.

Most corrosion rate data published in the literature is in empirical units. Ipy is equal to 25 millimetres per year in SI units. It is important to remember that corrosion rates expressed in millimetres per day (mdd) are dependent on material density. A 100 mdd is equal to 0.02 ipy for ferrous metals.

5.6 Acceptable Corrosion Rate of Materials

What are the parameters that will determine the acceptable corrosion rate of materials? Before we can determine the acceptable corrosion rate of materials, we need to know what they are.

The corrosion rate is largely determined by the duty, the economic life of the plant, the cost of the material, and the safety of the environment. The table below shows the acceptability criteria for the more commonly used inexpensive materials, such as carbon and low alloy steels.

Divide the values provided in table 1 by two for more expensive alloys, such as high alloy steels, brasses, and aluminium.

Corrosion rate depends on temperature and concentration of corrosion fluid. Corrosion usually increases with an increase in temperature. Corrosion rates will also be affected by other factors that are also affected by temperature, such as oxygen solubility.

Concentrations of corrosive medium also have complex effects. In diluted sulphuric acid, mild steel corrosion rates are more than 0.06 ipy, which is unacceptable. However, in concentrated sulphuric acid, corrosion rates above 70 per cent are acceptable.

Table 1. Acceptable corrosion rates

	Corrosion rate (ipy)	mm/y
Completely satisfied	< 0.01	0.25
Use with caution	<0.03	0.75
Use only for short exposures	<0.06	1.5
Completely unsatisfactory	> 0.06	1.5

Inhibition of Seawater Steel Corrosion via Colloid Formation

Nelson Cheng,[1] James Cheng,[2] Benjamín Valdez Salas,[3] M. Schorr,[3] and J. M. Bastidas[4]

[1]Magna International Pte Ltd., 10H Enterprise Road, Singapore 629834
[2]Magna Chemical Canada Inc., 1450 Government Road West, Kirkland Lake, Ontario, P2N 2E9, Canada
[3]Universidad Autónoma de Baja California (UABC), Mexicali, Baja California, Mexico
[4]National Centre for Metallurgical Research, CSIC, Spain

6.1 Abstract

The performance of a volatile corrosion inhibitor (VCI) on steel via colloid formation through its reaction with Ca and Mg ions in seawater was studied. The physical and chemical properties of seawater, with and without the VCI at different concentrations, were determined. The VCI's efficiency was assessed, and its suitability for the steel system in seawater was indicated at an optimal concentration of 0.05 per cent.

6.2 Introduction

Corrosion and degradation of materials are pernicious problems that affect environment quality, industry efficiency, and infrastructure assets.[1-2] All of these diverse facilities and installations require products, methods, and techniques to protect against, mitigate, and prevent corrosion damage. Volatile corrosion inhibitors (VCIs) are one of the modern technologies used to manage corrosion for the benefit of the global economy.[3]

6.3 Seawater Corrosion

The sea is a dynamic system in permanent motion. Complex surface currents and winds blowing over its surface generate waves that reach the coast and its industrial facilities located there.

Seawater is a solution consisting of many salts and numerous organic and inorganic particles in suspension. Its main characteristics are salinity and chlorinity, and from the corrosion point of view, dissolved oxygen (DO) content that ranges from 4 to 8 mg/L depending on temperature and depth. Seawater's minor components include dissolved gases—carbon dioxide (CO_2), ammonia (NH_3), and hydrogen sulphide (H_2S)—from urban sewage contamination. The oceans house algae, bacteria, and phytoplankton that generate about half of the oxygen in the atmosphere.

Ocean surface salinity is determined by the balance between water lost from evaporation and water gained through precipitation. The salt concentration, particularly sodium chloride (NaCl), varies from 2.0 to 3.5 per cent according to the sea location and added amounts of fresh river water. For instance, salinity of the Red

Sea (an enclosed basin) at high summer temperatures is 4.1 per cent, but salinity of the Baltic Sea is ~2.0 per cent since many rivers feed into it.

Seawater is slightly alkaline, with a pH of ~8. When it is contaminated by acids (i.e., in coastal regions near power stations burning fossil fuels and generating acidic rains), the pH can drop to 6.

6.4 Corrosion Inhibitors

In recent years, the use of VCIs has rapidly expanded worldwide for numerous technological and industrial applications such as cooling water systems;[4] steel-reinforced concrete; protected storage of military and electronic equipment;[5] acid pickling and cleaning;[6] the oil and gas industry, as additives to coatings, paints, and elastomers; and for corrosion avoidance in oil pipelines.[7-8] The importance and relevance of VCI technologies are evident by the many patents gathered in a recently published review.[9]

VCIs slow the rate of corrosion reactions when added in relatively small amounts to water. They are classified into three groups:

- anodic inhibitors, which retard the anodic corrosion reactions by forming passive films
- cathodic inhibitors, which repress the corrosion reaction (e.g., by reducing DO)
- adsorption inhibitors, such as amines, oils, and waxes, which are adsorbed on the steel surface to form a thin protective film that prevents metal dissolution

6.5 A Colloidal Corrosion Inhibitor

A polymolecular VCI, VAPPRO 844, was studied, which is added to seawater as a powder, and then it converts into a colloidal suspension with nanoparticles dispersed in the water. These nanoparticles are adsorbed on the steel surfaces, and a thin protective film is formed. The performance of this inhibitor depends on physical, biological, and chemical factors. The factors under analysis for this study included solution hardness, alkalinity, conductivity, and pH. Other factors, such as DO, contribute as well but were not within the scope of this investigation.

It is proposed that the mechanism of colloidal formation functions by combining the inhibitor (CI) with Ca^{2+} ions present in seawater to form an inert colloidal particle that is cationic in nature, as shown in equation (1):

$$Ca^{2+} + CI \rightarrow Ca^{2+}\text{-}CI \text{ complex.} \qquad (1)$$

The formed colloidal particles adhere to the metal and prevent the onset of corrosion by preventing the loss of electrons. This causes the electrochemical cell to be incomplete and corrosion cannot occur. The VCI powder was specially developed to combat corrosion on mild steel and iron structures in stagnant seawater found in ballast tanks of ships and rigs. In this study, the VCI was tested to establish its effectiveness and to determine the changes in both physical and chemical properties of the seawater, which include pH, total hardness, alkalinity, total dissolved solids, and conductivity at different VCI concentrations. The purpose was to find the optimum VCI concentration and provide recommendations on how the effectiveness of the inhibitor could be improved to reduce corrosion.

6.6 Results and Discussion

6.6.1 Weight Loss

The practices recommended in ASTM G3110 and NACE TM016911 were followed for evaluating the steel corrosion resistance. The measured weights for mild steel show that at 0.05 per cent concentration, there was the least weight loss, indicating the least corrosion. Over the period of twenty-six days, the steel control specimen in seawater without inhibitor had lost 0.58 g, while those specimens in seawater with inhibitor had reduced metal loss—~0.10 g on average. This was even lower than the tap water control of 0.15 g metal loss. The most effective VCI concentration was 0.05 per cent, as the metal loss was only 0.03 g (table 1).

The inhibition efficiency (IE) was determined using equation (2):

$$IE\% = \frac{M_u - M_i}{M_u} \times 100 \qquad (2)$$

where M_u and M_i are the weight loss of the steel in uninhibited and inhibited solutions, respectively.

6.6.1 Mild Steel Corrosion Reactions

A drop in solution hardness was observed; however, this was not reflected in the conductivity. This means that ions other than Ca^{2+} and Mg^{2+} had interacted in the seawater. The proposed reactions are shown in equations (3) and (4):

$$Cl + Ca^{2+}/Mg^{2+} \rightarrow \text{Gelatinous white precipitate} \qquad (3)$$

$$Cl + Ca^{2+}/Mg^{2+} + Fe^{2+}/Fe^{3+} \rightarrow \text{Insoluble complex} \qquad (4)$$

Table 1. Inhibition efficiency of VCI in seawater

Inhibitor Concentration (%)	Metal Loss (g)	Inhibition Efficiency (%)
—	0.58	—
0.0125	0.19	22.6
0.025	0.11	81.0
0.05	0.03	94.8
0.10	0.05	91.3
0.25	0.09	84.4

As iron underwent the anodic reaction in equation (5), the cathodic reaction expressed the oxygen reduction reaction under acidic conditions shown in equation (6) and under neutral alkaline conditions in equation (7):

$$Fe^{2+} \rightarrow Fe^{3+} + e^- \qquad (5)$$

$$O_2 + 4H^+ + 4e^- \rightarrow 2H_2O \qquad (6)$$

$$O_2 + 2H_2O + 4e^- \rightarrow 4OH^- \qquad (7)$$

In all of these reactions, the reduction of the hydrogen ions or the production of hydroxyl ions raised the pH of the electrolyte in fresh water. However, in seawater, the cathodic reduction observed by equations (8) and (9) produced a slightly alkaline surface condition, which precipitated calcium carbonate ($CaCO_3$) and magnesium hydroxide [$Mg(OH)_2$]:

$$Ca^{2+} + HCO_3^- + OH^- \rightarrow H_2O + CaCO_3 \qquad (8)$$

$$Mg^{2+} + 2OH^- \rightarrow Mg(OH)_2 \qquad (9)$$

On mild steel pieces in seawater with 0.25 and 0.10 per cent VCI and a pH range of 5 to 6, dark pits were observed on the metal toward the end of the analysis. These pits were much more likely to be formed at the anodic area because of the formation of the precipitate layer.

Steel pieces in seawater with 0.025 per cent or less VCI and a pH of 7.5 to 8.0 started to corrode. Thus, the inhibitor was not beneficial at such low concentrations.

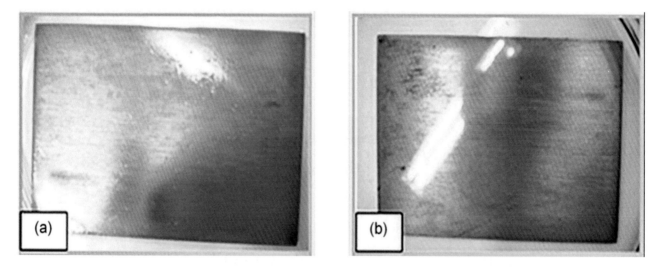

Figure 1. Mild steel samples exposed to seawater and 0.05 per cent VCI 844 before (a) and after (b) immersion in ferroxyl indicator.

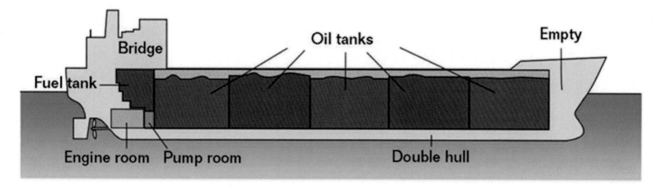

Figure 2. Petroleum transportation tanker showing holds.

With 0.05 per cent VCI in seawater, the pH range was ~7. Immersion in a ferroxyl indicator and weight-loss tests demonstrated that there was optimum corrosion inhibition at this concentration, although the metal had some staining (figure 1).

6.7 Applications

About 4,400 petroleum transportation tankers from oil-producing countries cross the oceans and seas of the world to energy-consuming countries. If, on average, each tanker has 10 holds, it means 44,000 holds require a VCI for their ballast seawater.

Petroleum steel tankers (figure 2) are cheaper and more efficient than submarine pipelines installed on the seabed for oil transportation. For their trip back, the tanker holds are filled with seawater to provide adequate stability (figure 3). A VCI is added to this ballast water. Pipes, storage tanks (figure 4), and pumps using water for hydrotesting also can be dosed with the same VCI.

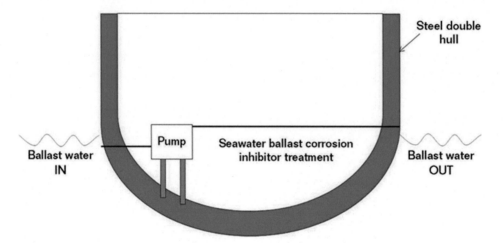

Figure 3. Ballast water tank.

Figure 4. Fire protection water storage tank.

6.8 Conclusions

From the experimental observations, mild steel was well protected with a VCI concentration of 0.05 per cent, showing only slight staining after a period of twenty-six days.

36

Changes in seawater parameters were observed when the VCI powder was introduced. It contributed to the increase of conductivity when introduced into the solution; however, when it reacted with the ions in seawater to form colloids, the conductivity dropped. The introduction of the VCI made the solution more acidic because of the mild acidic properties of this particular VCI.

Higher concentrations of inhibitor reduced the alkalinity of the seawater. For solution hardness, the calcium and magnesium ions were consumed in the reaction. This confirmed that the VCI powder followed the proposed reaction mechanism to form colloids.

6.9 References

1. R. Hummel, *Alternative Futures for Corrosion and Degradation Research* (Arlington, VA: Potomac Institute Press, 2014), pp. 2–13.
2. R. Raichev et al., *Corrosión de Metales y Degradación de Materiales*, M. Schorr, ed. (Mexicali, Baja California, Mexico: Universidad Autónoma de Baja California, 2009), pp. 281–284.
3. R. Garcia et al., 'Green Corrosion Inhibitor for Water Systems', *MP* 52, 6 (2013): pp. 48–51.
4. M. Schorr et al., 'Materials and Corrosion Control in Desalination Plants', *MP* 51, 5 (2012): pp. 56–61.
5. B. Valdez et al., 'Application of Vapour Phase Corrosion Inhibitors for Silver Corrosion Control in the Electronic Industry', *Corrosion Reviews* 21, 5-6 (2003): pp. 445–457.
6. Carrillo et al., 'Inorganic Inhibitors Mixture for Control of Galvanic Corrosion of Metals Cleaning Processes in Industry', *Corrosion* 2012 (Houston, TX: NACE International, 2012).
7. J. Hilleary and J. Dewitt, 'Corrosion Rate Monitoring in Pipeline Casings', *MP* 53, 3 (2014): p. 28.
8. T. Murthy, 'Monitoring of Chemical Treatment is Essential to Prevent Internal Corrosion', *MP* 53, 9 (2014): p. 54.
9. R. G. Inzunza et al., 'Corrosion Inhibitors Patents for Industrial Applications—A Review', *Recent Patents on Corrosion Science* 3, 2 2013): pp. 71–78.
10. 1ASTM G31-13, 'Standard Practice for Laboratory Immersion Corrosion Testing of Metals' (West Conshohocken, PA: ASTM International, 2013).
11. NACE TM0169-2000, 'Laboratory Corrosion Testing of Metals' (Houston, TX: NACE, 2012).

Technological Applications of Volatile Corrosion Inhibitors

Nelson Cheng,[1] Benjamín Valdez Salas,[2] Michael Schorr,[2] Ernesto Beltran,[2] and Ricardo Salinas[2]

[1]Magna International Pte Ltd., 10H Enterprise Road, Singapore 629834
[2]Universidad Autónoma de Baja California (UABC), Mexicali, Baja California, Mexico

7.1 Abstract

The objective of this review is to create a body of knowledge on the theoretical and practical aspects of corrosion inhibition to prevent and/or to eliminate corrosion in natural environments such as water, air, and acids and in industrial facilities such as oil, natural gas, concrete, paints and coatings, electronics, and military equipment. Corrosion inhibitors (CIs) and volatile corrosion inhibitors (VCIs) are applied in diverse forms such as powders, pellets, aqueous, or solvent solutions and in impregnated papers; closed in pouches and sachets; and added to coatings. Natural CIs are extracted by water or organic solvents from suitable plants. They represent the advanced trends of corrosion management based on green chemistry.

Keywords: corrosion, corrosion inhibitors, environments, industries

7.2 Introduction

Inhibition, in general, is a process aimed to restrain an activity in natural and industrial areas. In medicine, inhibitors arrest the action of an organ or a tissue in the human body (e.g., they bind to enzymes to decrease activity). Recommendations for the application of norms of inhibition are recorded in psychology manuals to improve human behaviour. In water, scale inhibitors maintain salts in solution to avert deposits of mineral scale. In this work, inhibitors used to prevent, avoid, or mitigate corrosion processes and events are reviewed.

Corrosion and pollution are pernicious problems that affect environment quality, industrial efficiency, and infrastructure assets (Raichev et al. 2009; Hummel 2014). Many pervasive pollutants produced by power stations burning fossil fuels accelerate corrosion, and corrosion products such as rust, oxides, and salts pollute bodies of water (Raichev et al. 2009; Valdez et al. 2012).

The aim of this review is to build a body of knowledge on the theoretical and practical aspects of corrosion inhibition, useful for the selection of VCIs to prevent and to eliminate corrosion in natural environments and industrial facilities.

The economic and social relevance of the corrosion management and control industry is evident in the activities of diverse international and national professional associations and R & D institutions dealing with all aspects of corrosion science, engineering, and technology such as the World Corrosion Organization (WCO), NACE International, the Worldwide Corrosion Authority, with its central office at the USA; the European Federation of Corrosion (EFC), CEBELCOR, Centre Belge d'Etude de la Corrosion; and many national organisations operating in industrial and developing countries. In the annual NACE conferences, the subject of CI is widely treated. For instance, in the 2016 conference, technical symposium research on corrosion inhibition by volatile corrosion inhibitors (VCIs), on corrosion control in oil and gas production with inhibitors, on coatings containing inhibitors, and on inhibitors for water reuse systems was presented. Furthermore, the importance of the dissemination of corrosion information is demonstrated by the numerous journals published in several languages. It is worthwhile to note that one journal is jointly dedicated to corrosion and scale inhibition. In this collection of *Corrosion* journals, it is appropriate to include the NACE International Corrosion Press, a newsletter that presents information on corrosion events and their curative treatment and offers solutions to current corrosion problems. Every five years, European Symposium on Corrosion Inhibitors is held at Ferrara, Italy, organised by the European Federation of Corrosion.

7.3 Volatile Corrosion Inhibitors

This review presents VCI as an economical and useful tool to control corrosion in environments such as water, air, soil, acids, and road de-icing; in industrial facilities such as oil, natural gas (NG), chemicals, concrete, and electronics; and in marine and prolonged protection of military equipment.

The use of VCI or vapour phase corrosion inhibitors (VPCI) has rapidly expanded in the last decades; in particular, a special type called VPCI was also designed as VCIs. VCIs slow the rate of corrosion reactions when added in relatively small amount to a corroding system. Corrosion inhibitors are classified into the following categories:

- anodic inhibitors, which retard the anodic corrosion reaction by forming passive films
- cathodic inhibitors, which suppress cathodic reaction, such as reduction of dissolved oxygen (DO)
- mixed inhibitors, which interact with both anodic and cathodic reactions
- adsorption inhibitors such as amines, oils, and waxes, adsorbed on the steel surface forming a thin protective film and preventing metal dissolution

VCIs contain organic and inorganic chemical compounds able to vaporise and condense in the presence of moisture-forming thin films on metallic surfaces. The development of these thin films and how the metallurgical and microstructural aspects of steel play a role in the corrosion mechanism are described in learned publications (Raja et al. 2003; Bastidas et al. 2005).

Sometimes, VCIs are impregnated in plastic bags and cover films or kraft wrapping paper, closed in pouches and sachets, or utilised in the form of powders or pellets. Kraft paper is made from chemical pulp that was produced through kraft process, a sulphate process that converts wood into pulp; this is the most widely used process. The process differs from normal paper production, for the solution used to convert wood into wood pulp consists of water, sodium hydroxide, and sodium sulphide. In addition, VCIs are also formulated in liquid aqueous or oil solutions, which can be sprayed over the metallic surfaces to protect. Some corrosion mechanisms focus on the thermodynamics and kinetics of the protection provided by the VCIs. The electrochemical nature of most corrosion processes requires similar mechanisms for the VCI performance during the corrosion inhibition. Ions or heteroatoms such us oxygen (O), sulphur (S), or nitrogen (N) contained in organic molecules such as amines, aldehydes, alcohols, carboxylic acids, and thiols, among others, and even π bonds enhance the adsorption of protective films on the metallic surfaces (Subramanian et al. 2000). Volatilisation capability of the inhibitor substance is one of the most important factors for an efficient corrosion inhibition. Volatile chemicals such as morpholine and hydrazine are added into boilers and transported by steam where they neutralise carbon dioxide or increase the pH rising alkaline value to protect condenser pipes from corrosion. In closed spaces, volatile cyclic amine salts are used as VCIs, as their vapour condenses and hydrolyses with the moisture, releasing protective ions that interact with the metal surface (Roberge 1999).

VCIs volatilise into the air and inhibit corrosion on metallic materials and metal-based products, particularly during shipment and storage. They are cost-effective and help conserve resources, and then they are incorporated into packaging materials (McConnell 2008).

The corrosion rates of reactive metals (e.g., Fe) are decreased by a modification of their surface with the mentioned organic molecules. This is the useful function of adsorbed VCIs. Their inhibition efficiency (IE) might reach 98 per cent, including degradable, non-toxic, natural VCI. The era of green VCI has already started (Costa and Marcus 2015).

Types of VCIs include nitrite of amines, amine carboxylates, heterocyclic compounds (thiazole, triazole, pyrrole, mercaptans, imidazoline, etc.), carboxylic acid esters, amines, acetylenic alcohols, and the mixtures or reaction products of these substances. Currently, the use of certain amines and nitrites has been prohibited by environmental and health regulations, as they can form nitrosamines, which can produce cancer. Figure 1 shows the structures of some typical VCIs.

VCI is a modern and economical technology for the reduction of corrosion. Its importance is evident by the patents gathered in a recently published review (Inzunza et al. 2013).

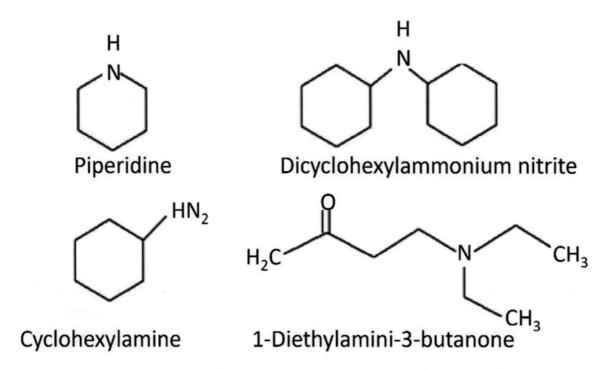

Figure 1. Chemical compounds used for the production of volatile corrosion inhibitors.

7.4 Green Corrosion Inhibitors

The advanced field of green chemistry, also known as sustainable chemistry, involves the design of chemical products and processes that reduce or eliminate the generation and use of hazardous substances as by-products. Sastri deals in his book *Green Corrosion Inhibitors: Theory and Practice* with CI adsorption on metal surface and its corrosion inhibition mechanism in different media: acid, neutral, and alkaline. The book presents guidelines for testing the toxicity, biodegradation, and bioaccumulation of CIs; standards for their environmental testing; and models to use in industrial practice. Sastri calls the environment-friendly CIs 'green', but toxic CIs are termed 'gray'. An additional classification of CIs considers them as hard, soft, and borderline (Sastri 2011).

Sharma has published an advanced book entitled *Green Corrosion Chemistry and Engineering: Opportunities and Challenges*, addressing the conflicts of societies and economies associated with corrosion problems and their real solutions and presenting an up-to-date overview of the progress in corrosion chemistry and engineering but emphasising the aspects of 'green' chemistry. Green chemistry technologies provide numerous benefits: safer products, depressed waste, saving critical resources such as water and energy, and improved chemical manufacture.

Two primordial chapters—'New Era of Eco-Friendly Corrosion Inhibitors' and 'Green Corrosion Inhibitors Status in Developing Countries'—are devoted to the fundamentals and application of CIs. The author deals with the diverse types of CIs such as anodic, cathodic, and mixed inhibitors; VCI; and precipitation inhibitors. Natural products mainly extracted from vegetables are considered green CIs. Sharma describes recent research and progress in these CIs and their uses in developing countries. Corrosion inhibition processes and CI practical utilisation are the core and the essence of this book. Furthermore, the book promotes the research for newer inhibitors for diverse applications (Sharma 2012). Sastri has focused his book on green VCIs, covering all aspects of basic principles and their modern application, with a full range of topics on the environment and industries as presented in this detailed review. He presents two types of green VCIs: based on amino acids and those extracted from natural plants. Other VCIs are produced to combat atmospheric corrosion (Sastri 2011).

Petroleum-based VCIs are replaced by VCIs obtained by extraction with solvents derived from vegetables such as soy and canola oils. These VCIs are biodegradable, contributing to maintain a healthy environment (Kharshan and Cracauer 2011).

Eco-friendly VCIs, also known as 'green' CIs, devoid of toxic components and biodegradable, are extracted from plants and presented in a large learned paper (Kesavan et al. 2012).

Another group of non-toxic and biodegradable VCIs has been proven effective in controlling the corrosion of steel, stainless steels, iron, aluminium and its alloys, cooper, and steel in concrete structures. This report provided a focused definition of sustainable development (Sharma et al. 2008).

VCI packaging in the worldwide market trend includes recycled kraft paper, which is reprocessed after use and has biodegradable and non-toxic properties.

7.5 Applications of Volatile Corrosion Inhibitors

7.5.1 Atmospheric Corrosion

Carbon steel is the main material employed for the fabrication of equipment that should be manufactured, transported, and stored under aggressive atmospheric conditions in tropical, arid, sandy, rainy, and humid regions.

Subramanian et al. (2000, 2002) report on the examination of the atmospheric corrosion resistance of steel, copper, and brass machinery supported by the application of octylamine, a VCI that has antibacterial properties. When equipment processes are wrapped with octylamine-impregnated kraft paper, the corrosion IE is 80 per cent for all three metals.

A special VCI is produced by impregnation of 20 per cent amino-carboxylate CI (ACCI) into kraft paper, obtained by applying a particular process for production of pulp paper. Experiments were carried out to assess the effectiveness of kraft paper, wrapping them around polished steel plates and placing them in a humidity chamber with 100 per cent humidity level for five days. Amino-carboxylate of 5 per cent has the best effect on corrosion protection (Valdez et al. 2017).

Kumar tested the efficiency of four VCIs regarding carbon steel under different atmospheric conditions at 40°C under high relative humidity containing 3 per cent NaCl. IE was in the following order: n-caprylic acid > n-butyric acid > 2-amino benzothiazole > N, N-dimethyl propylene urea (Kumar et al. 2013).

Temporary protection of carbon steel surfaces during transport and storage was achieved by applying a VCI, bio-piperidinium-menthol-urea (BPMU). Atomic force microscopy (AFM) examination showed that the interaction between the steel surfaces and VCIs generated an adscription protective film (Zhang et al. 2006).

Thermally stable VCIs absorbed on porous inorganic substrates, such as zeolites and diatomaceous earth, provided protection against corrosion during transport and storage. The VCIs studied were dicyclohexylammonium p-nitrobenzoate and phosphate. These VCIs reduce the corrosion rate (Estevao and Nascimiento 2001).

7.5.2 Water Supply

Fresh water comes from rain and snow. It is gathered in lakes and rivers, which are the sources of potable water for human consumption. They contain total dissolved solids (TDS) including chlorides, sulphates, and phosphates. Water is conveyed by pipelines made mainly of carbon steel, which suffer from corrosion. Water quality and its influence on human health depend on the pipeline performance and whether it is free from corrosion, scaling, and fouling. CIs are applied in sectors of the water industry such as cooling waters in power stations, desalination plants, potabilisation plants, wastewater reuse facilities, pollutants and microorganisms removal, and after sanitation and disinfection operations.

The CIs applied in several industrial cooling water systems are shown in table 1 (Sastri 2011; Garcia et al. 2013).

7.6 Acid Corrosion Inhibition

A great variety of acids, sulphuric, nitric, hydrochloric, hydrofluoric, phosphoric, acetic, sulphamic, citric, and formic, are broadly applied in environments and industries: water purification, fertilisers production, mines excavation, mineral leaching, food production and processing, metals cleaning and pickling and acidising petroleum wells and refineries distillation, fracturing of shale gas matrix, and more (Schorr and Valdez 2016). The acids undergo ionic dissociation in water, forming hydronium ion and an anion:

$$HA + H_2O \rightarrow H_3O^+ + A^-. \tag{1}$$

The acids attack carbon steel:

$$Fe + 2H^+ \rightarrow Fe^{+2} + H_2. \tag{2}$$

Well-acidising procedures are required in the oil, gas, and geothermal drilling industry to improve productivity. HCl is employed at various concentrations (depending on the reservoir characteristics), leading to acidic corrosion of steel, thus effective CIs are employed such as acetylenic and propargyl alcohols. Surfactant chemicals are added to assist in the formation of protective adsorbed films (Jackson 2016).

HCl and H_3PO_3 are the main acids selected for pickling of steel equipment such as body cars in the automotive industry. The CIs used are mixtures of nitrogen-bearing organic compounds, acetylenic alcohols, and sulphur-containing organic molecules. Commercial CIs were available and dissolved in water or organic solvents (Sastri 2011).

Innovative research describes the use of Schiff base compounds as CI for carbon steel in acid media. Quantum chemical calculations were conducted to determine the relationship between IE and the molecular structure of the CI. The Schiff base molecules were adsorbed on the steel surface providing corrosion resistance (Ju et al. 2017; Jw et al., 2017).

Table 1. Corrosion inhibitors for cooling waters

Cooling Water	Corrosion Inhibitors
Engine coolants	– molybdate
	– molybdate with nitrite; molybdate, arsenite, or arsenate and benzotriazole along with borate/phosphate/amine
	– Nitrite, nitrate, phosphate, borate, silicate, benzoate, aminophosphonate, phosphinopolycarboxylate, polyacrylate, hydroxybenzoate, phthalate, adipate, benzotriazole, tolyltriazole, mercaptobenzothiazole, and triethanolamine are combined with molybdate. In glycol, 0.1–0.6 wt% of molybdate is used.
Closed recirculating cooling water	– 200 ppm sodium molybdate with 100 ppm of sodium nitrite
	– 50 ppm molybdate, 50 ppm phosphate, 2 ppm Zn^{2+}
	– 40 ppm sodium molybdate + 40 ppm sodium silicate
	– 2-phosphonobutane-1,2,4-tricarboxilic acid and polyvinylpyrrolidone

Cooling Water	Corrosion Inhibitors
Cooling water of steam plant boiler waters	– molybdate with an aluminium salt and thiourea – mild steel corrosion inhibition in boilers by a mixture of sodium molybdate, sodium citrate, manganese sulphate, polymaleic acid, and morpholine – protection of mild steel in hard water boilers by sodium molybdate and sodium nitrite

The corrosivity of industrial phosphoric acid (PA) solutions of different concentrations depends on the chloride and the fluoride contents and the conditions of their use. Various investigators have proposed to employ organic compounds as CI to protect steel equipment in PA solutions, as follows: acid extracts of piper guanines, by Oguzie (2012); piperidine derivatives, by Ousslim (2014); N-containing organic compounds, by Zarrouk (2013); and benzyltrimethylammonium iodide, by Li et al. (2011). Full details are given in their articles listed in the references previously mentioned.

7.7 Corrosion Inhibitors Extracted from Vegetables

In the past, people living around jungles and forests discovered and extracted active biomaterials from trees and shrubs and used them to prevent and cure diseases and to relieve pain, improving the quality of their lives. Nowadays, the pharmaceutical enterprises, with their sophisticated chemical laboratories, synthesise their drugs and remedies, sometimes similarly to natural medicines.

Following the actual trend of 'green' chemistry, natural CIs are extracted from suitable plants, mixed, and applied in corrosion control industry. Numerous commercial vegetables (e.g., tobacco) are involved in the production of these novel CIs.

The tobacco industry comprises many products: dried tobacco leaves, stems, dust, liquid extracts. The use as CIs is based in their addition to steel coating, to increase their ability to protect against corrosion. Many countries grow tobacco; the residues are considered for anticorrosion activity (Von 2008).

Inulin is an extract from the roots of the chicory plant, used as a particulate CI mixed with an aqueous acid solution; the fluid is introduced into a well or supplied to a pipeline. The particle size of inulin is in the range of that of powder. It may be suspended in liquid for ease of handling (e.g., water or oil), depending on its application. The concentration range of inulin is from 0.01 per cent wt/vol to 5 per cent wt/vol of the aqueous acid solution (Choudary et al. 2012).

Laboratory investigations were performed to assess the use of aqueous and ethanolic extracts from plants encountered in the arid regions of the state of Baja California, denominated as creosote bush (*Larrea tridentata*) and *Pachycormus discolour*, in HCl solutions. The inhibition action of ethanol extract of the creosote bush in HCl aqueous was evaluated and found to be effective on carbon steel surfaces. HCl is employed for pickling and also in removing carbonate scales from steel surfaces (Inzunza et al. 2013; Inzunza 2014).

7.8 Electronics Industry

The worldwide electronics semiconductor industry produces microelectronics devices essential for the manufacture of digitally controlled televisions, telephony, radio, transmission equipment, instruments, vehicles, aircraft, and watercraft.

Advanced sensors and actuators are installed in autonomous vehicles, robotics, smart homes, machines, instruments, satellites, etc. Today, there is no human activity, industrial production, energy generation, water supply, transportation and communication, infrastructure maintenance, health care, or economy management that is not provided with electronic equipment.

Many metals and alloys, silver, copper, gold, aluminium, are applied in the fabrication of these devices. The indoor environment of manufacturing plants requires the control of conditioned air by the use of EPA filters to avoid penetration of corrosive pollutants. VCI sprays, emitters, or impregnated plastic and paper films are employed to protect against corrosion. The main pollutant is hydrogen sulphide (H_2S), which tarnishes silver devices, producing silver sulphide, and altering the electronic behaviour of the components. The combined use of chemical filters and VCIs improved the corrosion protection of copper and silver components in microchips manufactured at clean room conditions in a semiconductor plant.

The use of filters to trap chemicals in a TV manufacturing plant located in Mexicali, Mexico, was not enough to avoid the inlet of H_2S, which raised concentration above the 100 ppb, causing corrosion on silver surfaces. To control this sliver corrosion, VCI, VAPPRO (Vapour-phase-protection) 870 (Magna International Pte Ltd., Singapore) (trade name), was sprayed on silver electronic devices. The VCI product prevents corrosion in silver components, which retain their electronic properties and the ability for wave soldering process (Valdez et al. 2003, 2012).

Chemical mechanical polishing (CMP), also called planarisation, is applied to remove materials from the surfaces of microelectronic device wafer. Several patents are available on CMP, which contain mixed CI and colloidal silica. Other CIs based on sarcosine are utilised to protect printed circuit boards (Saji 2010).

7.9 Petroleum Industry

Petroleum is a mixture of hydrocarbons, extracted together with water, salts, and gases from wells. Water contains corrosive agents: CO_2, H_2S organic acids, chlorides, and sulphates. Wells with H_2S are called sour wells, and sweet wells are those with CO_2 (Sastri 2011). The petroleum industry consists of a complex system of operations, starting with crude oil extraction from wells, onshore and offshore; its conveyance by pipelines to treatment plants; and its refining to produce many oil derivatives such as gas, gasoline, kerosene, naphtha, and asphalt (Garcia 2015).

The applied CI includes organic compounds containing nitrogen, such as aliphatic salty acid derivatives, imidazole, and quaternary ammonium compounds.

Oil refineries apply fractional distillation to produce diverse derivatives, with gasoline as the most valuable. Wet corrosion is controlled by applying passivating, neutralising, and adsorption types of VCI. During oil refining, corrosion augments because of the formation of HCl, H_2S, CO_2, and O_2. The acids are neutralised by the addition of alkali to reach a pH level of 7.0–7.5 (Saji 2010; Inzunza et al. 2013).

Gasoline contaminated with water causes corrosion in transport systems and motor vehicles, which is treated with VCIs formulated with esters of carboxylic or phosphoric acid, following the regulation in accordance with NACE TM-01-72 and ASTM D6651 (Sastri 2011).

Corrosion control of the carbon legs of an offshore oil platform was affected using VCI as powder, contained in a string of closed pouches, suspended in hangers in the space adjacent to the legs (Al-Sayed et al. 2014).

Crude oil is transported from the producing countries to the consuming countries in petroleum steel tankers. They are cheaper and more efficient than submarine pipelines. In their trip back, the tanker holds are filled with seawater to provide adequate stability. A VCI, VAPPRO 844 (trade name), is added to seawater as powder, which is converted into a colloidal suspension of nanoparticles dispersed in the water and adsorbed on the steel surface, forming a thin protective film (Cheng et al. 2016).

7.10 Natural Gas Industry

NG is a source of energy for industrial, residential, commercial, and electrical applications; it is also a source of raw material for the polymers and plastic industry. The sectors of the NG industry are drilling, production, storage, transportation, and distribution; all suffer from corrosion. NG is extracted from land and marine wells, containing salty and briny water and corrosive gases H_2S and CO_2 (Schorr et al. 2006).

The main engineering materials for construction of NG industry facilities are steel and concrete, for ports, wells, pipelines, marine platform, liquefied NG regasification plants and storage plants. Corrosion control is managed by applying technical processes for selection or corrosion-resistant material for equipment and facilities construction, protective paints and coatings, cathodic protection, and VCI. Many types of CIs are used: anodic, cathodic, film forming, scavengers for annihilation of H_2S or oxygen (Valdez et al. 2015).

Multicomponent mixtures of CI and amine carboxylates–based VCIs are mixtures of inhibitors employed in NG depending on the characteristics of the NG systems, water content, and operation temperature (Sastri 2011; Inzunza et al. 2013; Garcia 2015).

The formation of black powder, constituted by corrosion powders, iron sulphides, ion oxides, and iron carbonates produced by corrosion attack of H_2S, CO_2, and O_2 in wet NG, has been detected in NG pipelines. Application of a combination of VCIs, more effective dehydration, and process control contributed to the solution of this problem (Olabisi 2017).

7.11 Concrete Corrosion

Concrete is a composite material, useful for structures in ports, airports, roads, bridges, stadia, etc. It is made of a mixture of water, portland cement, sand, and mineral aggregates. Steel reinforced concrete is generally very durable; nevertheless, concrete infrastructure—in particular, in marine environments—can undergo visible damage because of penetration of seawater, reaching the steel reinforcement. Concrete structures require expensive maintenance programs, which include cathodic protection, paints and coatings; therefore, CIs and VCIs are considered appropriate protection alternatives. Accordingly, VCIs are inserted into concrete structures, either during construction or after their finishing. Preferred VCIs are amine carboxylates cyclohexylamine nitrites, benzonates, and carbonates (Inzunza et al. 2013).

Calcium nitrite is a widely used VCI in concrete as it provides protection in the presence of chlorides; it does not affect the properties of concrete and is available for commercial utilisation. $Ca(NO_2)_2$ promotes the formation of a protective oxide film on the steel rods. Although nitrites have a good performance as VCIs, their use is regulated by environmental laws, and VCI formulations must be designed nitrite-free. Other CIs that have been found effective are benzoates, molybdates, borates, amines, and esters (Sastri 2011).

During harsh winters in the north-eastern states of USA and the north European countries, plentiful snow falls on roads and highways, freezing into ice. Salts and chemicals are spread on the snow to decrease the freezing temperature, to prevent freeze-thaw cycles, and to avoid road accidents. These corrosive de-icing chloride salts damage the road steel-reinforced concrete and corrode the metallic parts of cars and trucks bodies. Corrosion inhibitors, artificial and natural (extracted from vegetables), are added to the de-icing products to prevent, avoid, and/or mitigate corrosion (Augst et al. 2016). The de-icers, NaCl, $CaCl_2$, $MgCl_2$, have an environmental impact as they might leach into the soil; become toxic to humans, animals, and plants; and reach water bodies. VCIs produced by Magna International Singapore have been evaluated to know about their performance in inhibiting corrosion when used as de-icing salts. The behaviour of these VCIs has been studied in laboratory simulated tests of road de-icing conditions (Salinas et al. 2017).

7.12 Military Equipment

The combat fields of modern wars, including the struggle against global terrorism, are localised in diverse, harsh regions: tropical, desert, artic, marine and urban, with varied weather conditions, which adversely affect the corrosion resistance of the equipment, weapons, and vehicles involved. A significant development for corrosion control in the military services is the establishment of a central institution to serve the US Armed Forces. The US Department of Defense Office of Corrosion Policy and Oversight maintains a website, CorrDefense.org, which features contents on corrosion and corrosion control of military facilities, equipment, and weapons with the active participation of NACE International and the support of the North Atlantic Treaty Organization and the National Aeronautical and Space Administration (*CorrDefense* electronic magazine, US Department of Defense, 2002).

Parts of the equipment are kept in closed paper or polymeric envelopes and films or in cardboard boxes or wrapped with cotton canvas, well impregnated with a VCI. Sometimes, the parts are also sprayed with a VCI, depending on their size and shape. For long-term storage of vehicles and weapons, polyethylene, vinyl, and canvas tarps wetted with a VCI are used as covers, which also provide protection from ultraviolet radiation, humidity, rain, snow, dust, and mould. The covers are easy to install, are resistant to climate factors, and are durable (Cheng et al. 2016).

VCIs act by slow ions release mechanisms interacting with humidity within a sealed airspace, vaporising volatile anticorrosive compounds, which are deposited on the metal surface in ionic form. If the container is opened and reclosed, the inhibitor continues protecting the military equipment (Miksic et al. 2004; Bastidas et al. 2005).

7.13 Coatings, Paints, and Films

Coatings and paints applied on a metal surface form a barrier that impedes the access and action of the environment corrosive factors and are old tools for anticorrosion protection. Today, CIs are integrated into the coatings and paints or are deposited as a thin protective film. Furthermore, 'smart' coatings provided with nanoparticles can release CI on demand, and then an electrical or mechanical control signal is applied to the coatings. Such coatings are utilised in the aircraft industry, where CIs and the conversion coatings are able to detect pitting corrosion initiation. Organic-inorganic hybrid composites are employed in this methodology (Saji 2010).

VAPPRO VBCI 830 (a trade name) is impregnated in a mineral stone paper that is utilised and settled on the metal surface. The mineral paper contains a mixture of $CaCO_3$, SiO_2 powder, and VCI. Unlike pulp-based paper, this stone paper is UV-resistant and antistatic. It combines corrosion protection and packaging (Magna International 2016).

VCIs form adsorption layers of different thickness for the protection of high-precision tools in electronic, radio, and electronic equipment, where some VCIs form nanosized layers. Copper is the most important material in those industries because of its high electrical and thermal conductivity. To inhibit copper corrosion, benzotriazole and heterocyclic derivative molecules were used, and their performance was based on the formation of an insoluble film on the copper surface (Andreev and Kuznetsov 2005; Dehaghani 2016).

Certain VCIs generate passivation films on steel—in particular, those able to vaporise and react with the steel surface. This mechanism was investigated by polarisation electrochemical methods and electrochemical impedance spectroscopy (EIS). Figure 2 illustrates the formation of a protective film that contains VCI molecules, closed with a paper film that carries VCI too. Carboxylates, amines, and azoles form a protective film in neutral and alkaline solution. The passivation mechanism strongly depends on the pH solution (Rammelt et al. 2009).

7.14 Evaluation of VCI Performance

The VCI performance is evaluated following practices and standard methods to determine vapour inhibitor ability (VIA). The most commonly used is 4031 described in the federal standard FED-STD-101. Some practices applied to evaluate VIA are based on the above-mentioned method like the NACE TM0208-2013 or the German test method TL 8135-002 (German Federal Armed Forces 1980; British Standards Institution Procedures 1999; Department of Defense 2002; NACE International 2013). There are many reports in the literature that can be applied for VCI performance evaluation techniques (table 2).

Table 2: Test methods for evaluation of VCI performance.

Test method	Specimens materials	Cleaning	Test chambers	Control chamber	Test solution Glycerine/water	Saturation period (h)	Heating time	Condensation time (h)	Cycles	Evaluation
FED-STD-101 Test Method 4031	Carbon steel QQ-S-698 condition 5	Described	3	1	26% by volume of glycerine in water	20 h(at 22 ± 3°C)		3 at room temperature, add cold test solution at 4.4°C	1	Visual

Test method	Specimens materials	Cleaning	Test chambers	Control chamber	Test solution Glycerine/ water	Saturation period (h)	Heating time	Condensation time (h)	Cycles	Evaluation
NACE TM0208-2013	Carbon steel UNS G10100 (or UNS G10180) or cooper ASTM B15210	FED-STD-101 Test Method 4031	3	1	26% by volume of glycerine in water	20 h(at 22 $\pm$ 3°C)	5–20 s in warm bath water at 50 $\pm$ 2°C	3 at room temperature after adding cold test solution at 0–2°C	1	Visual
German Test Method TL 8135-002	Carbon steel DIN EN 10025	Described	3	1	26% by volume of glycerine in water, DIN 50008-1	20 $\pm$ 0.5 h (at 23 $\pm$ 2°C) without test solution. Test solution is added to the flask and stored for 2 h 10 m.	The set is kept for 2 h 10 m in a heating chamber at 40 $\pm$ 1°C.	Not applicable	1	NACE TM0208 2013
BSI* IEC 68-2-30:1980	-	-	-	-	-	-	40 or 55°C for 12 h	25 $\pm$ 3°C for 12 h at 95% RH	2, 6, 12, 21, 56 for 40°C. 1, 2, 6 for 55°C.	-

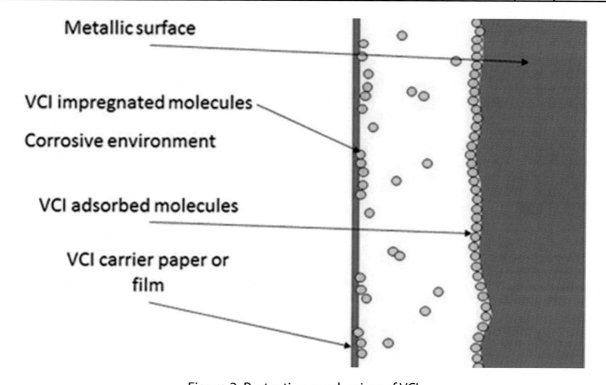

Figure 2. Protection mechanism of VCI.

Figure 3. Experimental set-up to evaluate VIA.

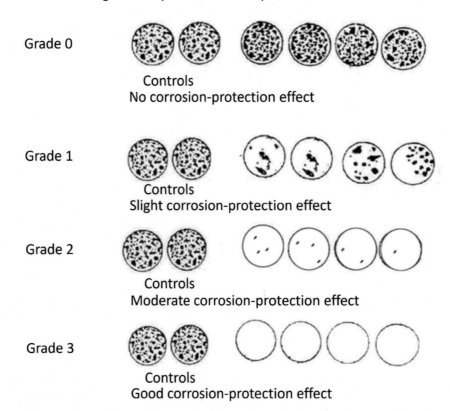

Grade 0

Controls
No corrosion-protection effect

Grade 1

Controls
Slight corrosion-protection effect

Grade 2

Controls
Moderate corrosion-protection effect

Grade 3

Controls
Good corrosion-protection effect

Figure 4. Visual patterns for rating the group of specimens according to NACE TM0208-2013.

Grade 4: Same as Grade 3, except when examined under 10× magnification, no more than three total spots among samples with no spot wider than 300 µm (0.012 in.). Excellent corrosion-protective effect.

All the tests were similar and performed according to the steps described as follows:

- Preparation of the metal specimens requires grinding by hand or machine. This preparation must yield polished test surfaces with highly reproducibility of the final finishing.
- Specimen cleaning process is performed following the recommendations of test 4031; each sample must be immersed in a tank or container of hot mineral spirits followed by immersion in methanol, allowed to dry in clean air, and then stored in a desiccator until ready to use.
- The test solution consists of a glycerine/water solution with a mass ratio of 1:2.
- Set-up of chamber for VCI materials includes the use of glass jars or Erlenmeyer flask depending on the selected test method (figure 3).
- VCI and moisture saturation period is one of the most important factors to be taken into consideration. During this time, the relative humidity (%RH) should increase to saturation of the closed container.
- Heating is sometimes required according to the procedure selected to perform the test. Some standards require rising of the temperature chamber.
- Condensation: After the saturation period, the temperature of the specimen is diminished to condense water on its surface and to form a CI film.
- Conditioning of specimens is the time in which the RH should increase and stabilise at a level greater than or equal to 90 per cent.
- Visual observation: after the specimen-conditioning period has elapsed, condensation of water should be visible, and corrosion should have occurred on the control specimens.
- Rating: Assign a numeral rating or grade to each metal specimen exposed in the test in accordance with the method selected. Visual patterns for rating the specimens according to NACE TM0208-2013 are displayed in figure 4.

7.15 Conclusions

The actual expansion of the economic, social, and military activities worldwide leads to the proliferation of corrosion phenomena and events, which should be combated.

Practical procedures that minimise or eliminate corrosion involve the selection of corrosion-resistant construction materials, application of coatings and linings, cathodic protection, and application of VCIs.

VCIs are broadly utilised for corrosion control in natural environments such as the atmosphere, water, soil, road de-icing, and industrial plants and facilities; electronics, petroleum, natural gas, concrete, coatings, and military equipment.

In the last decades, the use of 'green' VCIs, which include natural vegetables extracted into aqueous and solvents solution, is expanding. They belong to the novel field of 'green' chemistry.

7.16 References

1. Al-Sayed T., Eid A., Al-Marzooqi M., Jason U. Protection of offshore platform caisson legs with a vapor corrosion inhibitor—a case study, NACE Corrosion Conference, Paper No. 4200, March 2014.
2. Andreev N., Kuznetsov Y. Physicochemical aspects of the action of volatile metal corrosion inhibitors. Russ Chem Rev 2005; 74: 685–695.

3. Augst U. M., Buchler M., Sclumpf J., Marazzani B., Bakalli M. Long-term field performance of an organic corrosion inhibitor for reinforced concrete. Mater Perform 2016; 55: 36–40.

4. Bastidas D. M., Cano E., Mora E. M. Volatile corrosion inhibitors: a review. Anti-Corrosion Methods Mater 2005; 52: 71–77.

5. British Standards Institution Procedures. IEC-68-2-30:1980 Basic environmental testing, 1999.

6. Cheng N., Valdez B., Schorr M., Salinas R., Bastidas J. M. Corrosion inhibitors for prolonged protection of military equipment and vehicles. Mater Perform 2016; 55: 54–57.

7. Cheng N., Cheng J., Valdez B., Schorr M., Bastidas J. M. Inhibition of seawater steel corrosion via colloid formation. Mater Perform 2016; 55: 48–51.

8. Choudary Y. K., Sabhapondit, Ranganathan D. Inulin as corrosion inhibitor, US patent 20120238479A1, 2012.

9. Costa D., Marcus P. Adsorption of organic inhibitor molecules on metal and oxidized surfaces studied by atomistic theoretical methods in molecular modelling of corrosion process, Scientific development and engineering applications, Taylor and Marcus P, editors. New Jersey, USA: John Wiley & Sons Inc., 2015.

10. Dehaghani H. Diffusion of 1,2,3-benzotriazole as volatile corrosion inhibitor through common polymer films using the molecular dynamics simulation method. J Macromol Sci 2016; 55: 310–318.

11. Department of Defense, FED-STD-101 Test Method No. 4031 Vapor inhibiting ability of VCI materials, 2002.

12. Estevao L., Nascimiento R. Modification in the volatility rate of volatile corrosion inhibitors by means of host-guest systems. Corros Sci 2001; 43: 1133–1153.

13. Garcia I. Development of a polymeric coating containing corrosion inhibitors for corrosion control in marine environments, PhD Thesis, University of Baja California, 2015 (Spanish).

14. Garcia R., Valdez B., Schorr M., Eliezer A. Green corrosion inhibitors for water systems. Mater Perform 2013; 52: 48–51.

15. German Federal Armed Forces, TL 8135-002 Testing of anti-corrosive effect of vci auxiliary packaging materials, 1980.

16. Hummel R. Alternative futures for corrosion and degradation research, Arlington, VA: Potomac Institute Press, 2014: 2–13.

17. Inzunza R. G. Steel corrosion inhibitors of natural extracts for acid environments, PhD Thesis, University of Baja California, Mexico, 2014 (Spanish).

18. Inzunza R. G., Valdez B., Schorr M. Corrosion inhibitor patents in industrial application—a review. Recent Patents Corros Sci 2013; 3: 71–78.

19. Jackson J. The investigation of surfactant type molecular in acid corrosion inhibitor formulation, NACE Corrosion Conference, TEG 094X, 2016.

20. Ju H., Li X., Cao N., Wang F., Liu Y., Li. Schiff base derivatives as corrosion inhibitors for carbon steel material in acid media: quantum chemical calculation. Corros Eng Sci Technol 2017; 52: 1–8.

21. Kesavan D., Gopiraman M., Sulochana N. Green inhibitors for corrosion of metals: a review. Chem Sci Rev Lett 2012; 1: 1–8, ISSN: 2278–6783.

22. Kharshan M., Cracauer C. Application for biodegradable vapor phase corrosion inhibitors. Mater Perform 2011; 50: 56–60.

23. Kumar H., Saini V., Yadav V. Study of vapour phase corrosion inhibitors for mild steel under different atmospheric conditions. Int J Eng Innovative Technol 2013; 3: 206–211.

24. Li X., Deng S., Fu H. Benzyltrimethylammonium iodide as a corrosion inhibitor for steel in phosphoric acid produced by dihydrate wet method process. Corros Sci 2011; 53: 664–670.

25. Magna International Company, Vappro VCBI 830, VCI Mineral Stone Paper, www.vapprovci.com. Accessed 2017.

26. McConnell R. Volatile corrosion inhibitors offer effective protection for processing and shipment of metal-based products. Met Finish 2008; 106: 23–27.

27. Miksic B., Boyle R., Wuertz. Efficacy of vapor phase corrosion inhibitor technology in manufacturing. Corrosion 2004; 60: 515–522.

28. NACE International, TM208-2013 Laboratory test to evaluate the vapor-inhibiting ability of volatile corrosion inhibitor materials for temporary protection of ferrous metal surface, 2013.

29. Oguzie E. E., Adindu C. B., Enenebeaku C. K., Ogukwe C. E., Chidiebere M. A., Oguzie K. L. Natural products for materials protection: mechanism of corrosion inhibition of mild steel by acid extracts of *Piper guineense*. J Phys Chem C 2012; 116: 13603–13615.

30. Olabisi O., Al-Sulaiman S., Jarragh A., Khuraibut Y., Mathew A. Black powder in export gas lines. Mater Perform 2017; 56: 50–54.

31. Ousslim A., Bekkouch K., Chetouani A., Abbaoui E., Hammouti B., Aouniti A., Bentiss F. Adsorption and corrosion inhibitive properties of piperidine derivatives on mild steel in phosphoric acid medium. Res Chem Intermed 2014; 40: 1201–1221.

32. Raichev R., Veleva L., Valdez B. Corrosion de Metales y Degradacion de Materiales, Schorr M. editor. Universidad Autonoma de Baja California, 2009: 281–284.

33. Raja P. B., Ismali M., Ghoreishiamiri S., Mirza J., Che M., Kakooei S., Rahim A. Reviews on corrosion inhibitors: a short view. Chem Eng Commun 2003: 203; 1145–1156.

34. Rammelt U., Koehler S., Reinhard G. Use of vapour phase corrosion inhibitors in packages for protecting mild steel against corrosion. Corros Sci 2009; 51: 921–925.

35. Roberge P. R. Corrosion inhibitors, in Handbook of Corrosion Engineering. McGraw-Hill: New York, NY, USA, 1999: 833–862.

36. Saji V. A review on recent patents in corrosion inhibitors. Recent Patents Corros Sci 2010; 2: 6–12.

37. Salinas R., Schorr M., Valdez B. Corrosion inhibitors for concrete road de-icing operations, Latincorr Conference, October 2016, Mexico City, Report in Mater Perform, 2017: 176–177.

38. Sastri V. S. Green corrosion inhibitors: theory and practice, New Jersey, USA: John Wiley & Sons Inc., 2011.

39. Schorr M., Valdez B. The phosphoric acid industry: equipment, materials and corrosion. Corros Rev 2016; 34: 85–102.

40. Schorr M., Valdez B., Quintero M. Effect of H_2S in polluted waters: a review. Corros Eng Sci Technol 2016; 41: 221–227.

41. Sharma S. K. Green corrosion chemistry and engineering, Weinheim, Germany: Wiley-VCH Verlag GmbH, 2012, pp. 430.

42. Sharma S. K., Mudhoo A., Khamis E., Jain G. Green corrosion inhibitors: an overview of recent research. J Corros Sci Eng 2008; 11. ISSN: 1466–8858.

43. Subramanian A., Natesan M., Muralidharan V. S., Balakrishnan K., Vasudeban T. An overview: vapor phase corrosion inhibitors. Corrosion 2000; 56: 144–155.

44. Subramanian A., Rathina R., Netesan M., Vasudevan T. The performance of VPI-coated paper for temporary corrosion prevention of metals. Anti-Corrosion Methods Mater 2002; 49: 354–636.

45. Valdez B., Cheng J., Flores F., Schorr M., Veleva L. Application of vapour phase corrosion inhibitor for silver corrosion control in the electronics industry. Corros Rev 2003; 21: 445–456.

46. Valdez B., Schorr M., Zlatev R., Carrillo M., Stoytcheva M., Alvarez L., Eliezer A., Rosas N. Corrosion control in industry. In: Valdez B., Schorr M., editors. Environmental and industrial corrosion—practical and theoretical aspects. Rijeka, Croatia: Intech, 2012: 19–54.

47. Valdez B., Schorr M., Bastidas J. M. The natural gas industry: equipment, materials and corrosion. Corros Rev 2015; 33: 175–185.

48. Valdez B., Cheng N., Salinas R., Cheng J., Schorr M. VCI impregnated in Kraft paper for humid and saline environments, 2017 (under revision).

49. Von F. J. Coatings including tobacco products as corrosion inhibitors, WO Patent 2008; 151028 A3.

50. Zarrouk A., Zarrok H., Salghi R., Hammouti B., Bentiss F., Touir R., Bouachrine M. Evaluation of N-containing organic compound as corrosion inhibitor for carbon steel in phosphoric acid. J Mater Environ Sci 2013; 4: 177–192.

51. Zhang D., An Z., Pan Q., Gao L., Zhou G. Volatile corrosion inhibitor for formation on carbon steel surface and its inhibition effect on the atmospheric corrosion of carbon steel. Appl Surface Sci 2006; 253: 1343–1348.

CHAPTER 8

Corrosion Inhibitors for Prolonged Protection of Military Equipment and Vehicles

Nelson Cheng,[1] Patrick Moe,[1] Benjamín Valdez Salas,[2] Michael Schorr,[2] Ernesto Beltran,[2] Ricardo Salinas,[2] and J. M. Bastidas[3]

[1]Magna International Pte Ltd., 10H Enterprise Road, Singapore 629834
[2]Universidad Autónoma de Baja California (UABC), Mexicali, Baja California, Mexico
[3]National Centre for Metallurgical Research, Spanish National Research Council, Madrid

8.1 Abstract

Military assets require the implementation of corrosion control and monitoring techniques—in particular, during long periods of storage. Plastic sheets and/or textile canvas, impregnated with vapour phase inhibitors, are used to cover weapons and vehicles. An electronic corrosion control monitoring system for covered vehicles and weapons is described.

The combat fields of modern wars, including the struggle against global terrorism, are localised in diverse, harsh regions: tropical, desert, arctic, marine, and urban, with varied weather conditions that adversely affect the corrosion resistance of the equipment, weapons, and vehicles involved.

Corrosion and degradation of military hardware occur by an interaction between the surface of a material and its environment, damaging the equipment, weapons, vehicles, and machinery.[1-2] Localised corrosion can occur on parts of this equipment that are prone to corrosion. Many corrosion types are known: intergranular, in microcrystalline grain boundaries; galvanic, between different metals electrically interconnected; crevice, at interfaces between parts; pitting, forming perforations; de-alloying, by leaching of a less noble metal from an alloy; fatigue, with cracks generated by mechanical stress; fretting, caused by two metallic surfaces rubbing against each other; erosion-corrosion, by the combined action of mechanical wear; and electrochemical corrosion.[1] To prevent and mitigate these forms of corrosion, military assets require the implementation of corrosion control methods and techniques, including corrosion inhibitors, particularly the new 'green' corrosion inhibitors, according to the characteristics of the corrosive environment.[3]

A significant development for corrosion control in the military services is the establishment of a central institution to serve the US armed forces. The US Department of Defense (DoD) Office of Corrosion Policy and Oversight (CPO) maintains a website, CorrDefense.org, that features content on corrosion and corrosion control of military facilities, equipment, and weapons. In addition, an alliance was created between the US DoD CPO and defence departments in the United Kingdom, Canada, France, Germany, New Zealand, and Australia that ensures the expansion of corrosion control efforts worldwide.

8.2 Corrosion Inhibitors

The use of corrosion inhibitors is rapidly expanding worldwide for numerous technological and industrial applications: in cooling water systems;[4] protected storage of military and electronic equipment;[5] acid pickling and cleaning;[6] the oil and gas industry;[7] as additives to coatings, paints, and elastomers; for corrosion avoidance in oil pipelines;[8-9] and in desalination plants.[10] The importance and relevance of the inhibitor technology are evident from the many patents gathered in published reviews.[11-12]

Corrosion inhibitors slow the rate of corrosion reactions when added in relatively small amounts to a treated system. They are classified into three groups:

- anodic inhibitors—which retard the anodic corrosion reactions by forming passive films
- cathodic inhibitors—which repress the corrosion reaction such as reduction of dissolved oxygen
- adsorption inhibitors—such as amines, oils, and waxes, which are adsorbed on the steel surface to form a thin, protective film that prevents metal dissolution

Volatile corrosion inhibitors (VCIs) are used to protect ferrous tools and delicate machinery parts against corrosion, where it is impractical to apply surface treatments. These parts are kept in closed paper envelopes, in cardboard boxes, or wrapped with cotton canvas, well impregnated with a VCI. Sometimes, the parts are also sprayed with a VCI, depending on their size and shape. Figures 1 through 3 show several pieces of critical military hardware protected against corrosion with VCIs.

Figure 1. Heavy cannons on wheels covered with rough canvas fabric impregnated with a VCI.

Figure 2. Wrapped military hardware on a wood pallet ready for moving.

Figure 3. Weapon parts wrapped with transparent plastic sheets containing a VCI.

VCIs are evaluated by NACE International TM0208,[13] including some types of VCI materials (e.g., paper), using low carbon steels (CS) as representative of the broad class of ferrous metals. This standard was prepared by NACE Task Group 215, 'Volatile Corrosion Inhibitors'.

8.3 Military Asset Protection

The specific conditions of protecting military equipment in high-humidity climates and the diversity of materials used in the construction of military equipment call for special quality VCI products to protect them from corrosion. The slightest occurrence of corrosion on arms and other equipment can cause changes in their technical performance, affect their reliability in practice, and result in loss of lives in times of war. VCIs do not significantly affect human health or combat readiness.

In practice, most of the military equipment and arms are continually kept in reserve, so it is of paramount importance to protect them from corrosion while maintaining their original technical properties. Trends in protecting such equipment have been based on reliability, easy application, quick mobilisation, and economic benefits.

VCIs have been developed to meet the above requirements. Materials used are organic chemicals containing a proprietary amine compound. This compound is non-toxic, biodegradable, and environmentally friendly. Through years of experience with several armed forces around the world, cutting-edge technologies have been developed to preserve all types of military equipment.

Military assets consist of fixed and mobile structures, including two principal groups: weapons systems for attack and defence operations, and vehicular assets for transportation of military personnel, their supplies, and materials.[14]

8.4 Ground Vehicles

Military services operate many types of armoured vehicles, wheeled and non-wheeled. Exposed to harsh environments, they are assembled from numerous metallic components, but mainly CS and some low-alloy steels. Therefore, they are subjected to general corrosion, particularly in humid environments. Because of the limited corrosion resistance of these steels, they are often protected against corrosion by the application of chemical conversion coatings.

The most commonly used vehicle is the high-mobility multipurpose wheeled vehicle, also known as the Humvee, particularly in the deserts of the Middle East. The complex interaction of chemical and climatic factors, such as hot afternoons and cold nights, with dew condensation, intense solar radiation, and salt-laden atmosphere as well as saline, brackish, and briny water, create a harsh corrosive environment. The military vehicles are damaged by corrosion, erosion, abrasion, and wear that impair vehicle mobility and long service life, requiring costly ongoing maintenance. Plastics, elastomers, and composites are generally unsuitable alternatives because they deteriorate by physicochemical mechanisms.

Polyethylene, vinyl, and canvas tarps, wetted with a VCI, are used as covers for long-term storage of vehicles. These covers also provide protection from ultraviolet radiation, humidity, rain, snow, dust, and mould. The covers are easy to install, are resistant to the climate factors, and are durable.

Corrosion control monitoring systems help to maintain a high level of operational readiness for preserved military equipment and vehicles by drastically reducing corrosion problems during storage. Early detection and resolution of corrosion problems not only help reduce maintenance costs but also provide the logical methodology to maintain operational readiness while minimising the need for repair or replacement during mobilisation. The system enables real-time monitoring of preserved military equipment. It consists of sensor controllers and corrosion sensors and is designed to be easily scalable according to the number of equipment packs or vehicles to be monitored (figure 4).

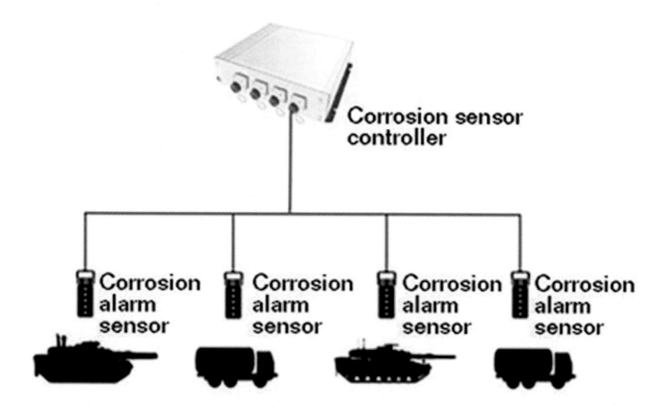

Figure 4. Corrosion control monitoring system.

Corrosion control monitoring software enables fleet management personnel to easily identify corrosion problems on a specific piece of equipment or vehicle. It alerts personnel upon detection of corrosion at a specific location, and the alert returns to its normal level only after the problem is rectified. This system reduces the regular need for on-site inspection and enhances productivity, helping to improve staffing allocation within the fleet management team.

8.5 Conclusion

Proper use of VCIs in combination with a system of automated corrosion sensors can be an effective means of preserving equipment and vehicles during long-term storage. Critical advantages include savings in maintenance costs and prevention of corrosion damage that can reduce efficiency.

8.6 References

1. R. Hummel, *Alternative Futures for Corrosion and Degradation Research* (Arlington, VA: Potomac Institute Press, 2014), pp. 2–13, 19–89, 56–57.

2. R. Raichev, L. Veleva, B. Valdez, *Corrosión de Metales y Degradación de Materiales,* M. Schorr, ed. (Mexicali, Mexico: Universidad Autónoma de Baja California, 2009), p. 281–284.

3. V. S. Sastri, *Green Corrosion Inhibitors: Theory and Practice,* 1st ed. (Chichester, UK: John Wiley & Sons, 2011).

4. R. Garcia, B. Valdez, M. Schorr, A. Eliezer, 'Green Corrosion Inhibitor for Water Systems', *MP* 52, 6 (2013): pp. 48–51.

5. B. Valdez, J. Flores, J. Cheng, M. Schorr, L. Veleva, 'Application of Vapour Phase Corrosion Inhibitors for Silver Corrosion Control in the Electronic Industry', *Corrosion Reviews* 21, 5-6 (2003): pp. 445–457.

6. I. Carrillo, B. Valdez, M. Schorr, R. Zlatev, 'Inorganic Inhibitors Mixture for Control of Galvanic Corrosion of Metals Cleaning Processes in Industry', CORROSION 2012 (Houston, TX: NACE International, 2012).

7. N. Cheng, J. Cheng, B. Valdez, M. Schorr, J. M. Bastidas, 'Inhibition of Seawater Steel Corrosion Via Colloid Formation', *MP* (under revision).

8. J. Hilleary, J. Dewitt, 'Corrosion Rate Monitoring in Pipeline Casings', *MP* 53, 3 (2014): p. 28.

9. T. Murthy, 'Monitoring of Chemical Treatment is Essential to Prevent Internal Corrosion', *MP* 53, 9 (2014): p. 54.

10. M. Schorr, B. Valdez, J. Ocampo, A. So, A. Eliezer, 'Materials and Corrosion Control in Desalination Plants', *MP* 51, 5 (2012): pp. 56–60.

11. R. G. Inzunza, B. Valdez, M. Schorr, 'Corrosion Inhibitors Patents for Industrial Applications, A Review', *Corros. Sci.* 3, 2 (2013): pp. 71–78.

12. D. M. Bastidas, E. Cano, M. Mora, 'Volatile Corrosion Inhibitors: A Review,' *Anti-Corrosion Methods and Materiels* 52, 7 (2005): pp. 71–77.

13. NACE TM0208-2013, 'Laboratory Test to Evaluate the Vapor-Inhibiting Ability of Volatile Corrosion Inhibitor Materials for Temporary Protection of Ferrous Metal Surfaces' (Houston, TX: NACE International, 2013).

14. I. C. Handy, J. Repp, 'Ground Vehicles Corrosion,' *ASM Handbook, Vol. 13C Corrosion: Environments and Industries* (Materials Park, OH: ASM International, 2006).

9

CHAPTER

Application of Vapour Phase Corrosion Inhibitors for Silver Corrosion Control in the Electronics Industry

Benjamín Valdez Salas,[1] Francisco Flores,[1] M.Schorr,[1] James Cheng,[2] and Lucien Veleva[3]

[1]Universidad Autónoma de Baja California (UABC), Mexicali, Baja California, Mexico
[2]Magna Chemical Canada Inc., 1450 Government Road West, Kirkland Lake, Ontario, P2N 2E9, Canada
[3]Proyecto TROPICORR—CYTED, CINVESTAV-IPN, Merida, Appl. Physics Dept., Carr. Ant. A Progreso, Km. 6, CP 97310, Merida, Yucatan, Mexico

9.1 Abstract

The indoor corrosion of silver components used in the manufacture of electronic devices represents a hard challenge for the electronic industry in Mexico. In this work, a case of silver corrosion occurring in a TV manufacturing plant was documented, analysed, and diagnosed. The main pollutant present in the indoor environment of the factory was hydrogen sulphide, which causes rapid silver tarnishing because of the formation of silver sulphide corrosion products. Silver corrosion rates were evaluated by gravimetric assays, and surface SEM and EDX analyses were performed to characterise the corrosion film. To control the corrosion process, VAPPRO vapour phase corrosion inhibitors were used.

Keywords: silver, electronic devices, hydrogen sulphide, VAPPRO VCI, TROPICORR

9.2 Introduction

Electronic equipment components, computers, integrated circuits (IC), and microchips in indoor atmospheres are exposed to a variety of environmental conditions, and, frequently, corrosion failure of these devices occurs. Corrosion is becoming an even more significant factor in the reliability of electrical and electronic equipment.[1] Within the last decade, the electronic and electrical industries are increasingly applying more vapour phase corrosion inhibitors (VCI) for electronic components and devices. As electronics continue to shrink in size and grow in capacity, the importance of corrosion control increases.[2, 3]

The interaction among electrical, metallurgical, and environmental conditions, together with severe dimensional constraints, presents a unique set of corrosion problems, which lead to malfunction, failure, and, finally, operation interruption. These problems appear in several industries and technologies applying electronic components and devices (e.g., communications, information, control, avionics, medical, meteorological, military, robotics, space, missiles, satellites, domestic, etc.).

Silver, copper, and gold are important functional materials found in electrical/electronic devices. New technologies, tools, laboratory methods, and instruments for failure analysis of electronic devices are being constantly introduced. The main objective of these activities is failure prevention.[4-7]

9.3 Environmental Conditions

The state of Baja California at the northwest of Mexico hosts more than 25 per cent of the electronic devices manufacturing industry. Corrosion of silver components occurs at indoor conditions in several industrial plants located at Mexicali City. The state capital Mexicali is a semi-desert zone irrigated by the Colorado River, which favoured the formation of the Mexicali Valley with extreme climatic conditions: hot in summer (48°C) and cold in winter (4°C). At twenty-five kilometres to the south are located the Cerro Prieto volcano and the second-largest geothermal field in the world, also named Cerro Priety. The geothermal vapour pollutes the atmosphere with hydrogen sulphide (H_2S) and other non-condensable gases.[8] The oxidation ponds for municipal wastewater treatment and the high quantity of automobiles circulating in the area are also considered as sources of sulphur compounds capable of polluting indoor installations and promoting corrosion of silver components.

Mexicali is a dusty city with strong winds that generate dust storms. Samples of Mexicali airborne dust, with a mean diameter of 10 μm (PM_{10}), are a mixture of clay minerals and quartz grains. Its chemical composition consists of 75 per cent potassium aluminium silicate and 20 per cent silica (SiO_2). These are hygroscopic mineral particles, facilitating the entry of humidity into the electronic devices affected by deposited dust particles.

The city of Mexicali, situated just on the border with the state of California, USA, applies the regulations of the National Institute of Ecology pertaining to the Mexican 'Secretaria de Medio Ambiente y Recursos Naturales'—SEMARNAT—for pollution control of the atmosphere. Several air-monitoring stations are installed and operated around Mexicali. On the other side of the international frontier, in Calexico and other nearby towns, the rules of EPA—the Environment Protection Agency—USA are enforced.

The corrosion of small and micro components of silver causes extensive losses to the electronics industry because of reject product, problems with soldering processes, and production delays. The silver surface becomes tarnished by the effect of H_2S, and, depending on exterior pollution, the problem can occur at controlled indoor conditions in clean rooms too. In this work, the causes of corrosion in silver electronic switches used in television manufacturing (figure 1) and the employment of VAPPRO VCI to prevent these noxious events were studied.

Figure 1. Electronic TV switches showing corrosion on silver contacts.

9.4 Experimental Procedures

The experimental sequence includes the characterisation of pollutants present in the indoor atmosphere, gravimetric corrosion tests on silver coupons at indoor installations, and surface analysis. The corrosivity of indoor conditions and the results of the corrosion tests were evaluated using ISO standards[9] recommended for the electronics industry.

Based on our experience and because of the Mexicali air quality characteristics, there is strong evidence that the coloured stains appearing on silver switches of the TV boards are due to corrosion by H_2S and formation of silver sulphide (Ag_2S). Several scientific papers have reported that a concentration of 100 ppb of H_2S in the indoor is enough to induce the build-up of sulphide films on the silver surfaces.[10-13]

9.5 Environment Pollution and Corrosivity

To identify the source of pollutants and the type of corrosion products, two approaches were followed:

- Monitoring of air pollutants: Two different sulphur compounds—H_2S and sulphur dioxide (SO_2)—in the range from 0 to 100 ppm, which are critical for silver corrosion, were monitored at different inside and outside locations of the plant using portable gas analysers. The gas analyser apparatus consists of transducers and electronic circuitry and uses chemiluminescence and infrared absorption methods to detect pollutant gases in the environment.
- Corrosion study on silver coupons: The environment corrosivity was determined by the evaluation of several parameters:
  - gravimetric corrosion tests on silver coupons by weight decrease or increase, in mg/m^2.year
  - indoor pollution by gases and variations of temperature and relative humidity (RH)

The silver coupons were made from metallic silver, 99.9 per cent purity, in a rectangular form: 50 mm length, 20 mm wide, and 1.1 mm thickness. Their surface was polished with SiC paper grade 120. The coupons were degreased with acetone vapour, weighed in an analytical balance, and kept in a dry place before their installation at indoor plant locations.

Indoor pollutants such as SO_2 and H_2S were monitored in the range from 0 to 100 ppm with portable electronic devices. Readings were taken from several places of the production area an hourly interval. A data logger–type device was used to record the relative humidity and temperature.

9.6 Surface Analysis

Several slightly corroded silver samples supplied by a TV manufacturing industry were initially analysed by scanning electron microscopy (SEM) and EDX without previous treatment to observe the corrosion products, morphology and their chemical composition. Also, some silver coupons exposed to corrosive environments were analysed using these techniques to examine corrosion products and particulate matter deposited on their surface. The particulate matter deposited, such as dust, is critical because it adversely affects the soldering on the silver surface, generating failures in the electrical contact with other electronic components.

9.7 Indoor Atmosphere Corrosion Categories

The corrosivity of indoor atmospheres was evaluated applying ISO/CD Standard 11844-A, which deals with metals, alloys, and metallic coatings subjected to atmospheric corrosion influenced by air humidity, pollutant gases, and solid substances. Corrosivity data are of fundamental importance for implementing suitable corrosion protection or for evaluating serviceability of metals in the electronics industry.

The evaluation of low-corrosivity indoor atmosphere is accomplished by direct determination of corrosion attack of relevant metals or by measurement of environmental parameters, which may cause corrosion of metals and alloys applied in the electronics industry.[7]

Indoors atmospheres considered in this study are classified into five corrosivity categories, denoted IC1 to IC5, as given in table 1. These categories are based on measurements of corrosion attack on standard metallic specimens after an exposure for one year. From the mass loss or mass increase, the indoor corrosivity category for each metal is determined.

Table 1. Corrosivity categories of indoor atmospheres according to ISO/CD 11844

IC1	very low indoor
1C2	low indoor
IC3	medium indoor
1C4	high indoor
IC5	very high indoor

9.8 Evaluation of Vapour Phase Corrosion Inhibitors

To control the corrosion problems, the application of VAPPRO VCI was recommended to the TV manufacturer. Its efficiency was evaluated at the indoor conditions in the plant. The VAPPRO 870 Electro-Spray was provided by Magna Chemical Canada Inc. (15 Bowman Ave., Box 534, Matheson, Ontario POK, INO, Canada). This company is devoted to the development of the most advanced and environmentally safe VCI technology, known as VAPPRO (Vapour-Phase-Protection). VAPPRO VCI technology comes in many forms, such as rust converters, VCI wax coatings and VCI grease, VCI emitters, VCI films, etc., and they are all effective for corrosion control. Magna Chemical Canada Inc. recommends using its VAPPRO 870 Electro-Spray product, a liquid solution composed of organic inhibitors in an isopropyl alcohol base. This specific VAPPRO VCI product is formulated for the galvanic corrosion protection of metals and alloys normally found in electronics applications such as copper, silver, aluminium, and other non-ferrous metals. For the purposes of this study, the VCI use must conform with several requirements—for instance, the VCI protective layer should not alter the thermal, electrical resistance, or magnetic properties of the metal. The effectiveness of VAPPRO VCI has been demonstrated in numerous field and laboratory tests.

To evaluate the inhibitor efficiency, duplicate specimens of silver and copper were placed in the most aggressive indoor zone. One set of specimens was placed in a semi-closed container where VAPPRO 870 Electro-Spray was sprayed to saturate the space with VCI. Another separate set of specimens of similar metals were placed nearby without any application of VAPPRO VCI for comparative purposes. The specimens were tested for ten days, and the surface conditions were evaluated by visual observation at the end of the test period. A soldering test using a rich tin lead-free solder was performed on both silver specimens.

9.9 Results

To carry out a diagnosis of the corrosion problems occurring, previous SEM and EDX assays on corroded silver samples were analysed. The results, shown in figure 2, indicate that these samples are covered with corrosion products consisting of silver sulphide and chloride. The silver specimen exposed in the plant for one month appeared with a dark, dirty surface; silica was found, probably because of the contamination with fine dust particles. Tarnishing of these silver specimens occurs after forty-eight hours of exposure because of the reaction with sulphur compounds present in the environment. The results of gravimetric corrosion tests performed in five different stations, four indoors and one outdoors, are shown in figure 3.

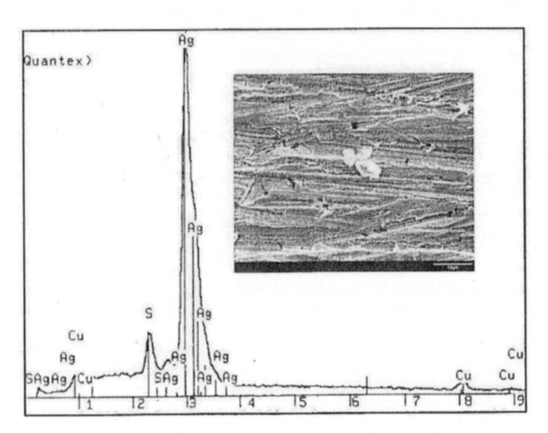

Figure 2. EDX spectra for corroded silver contacts on electronic TV switches.

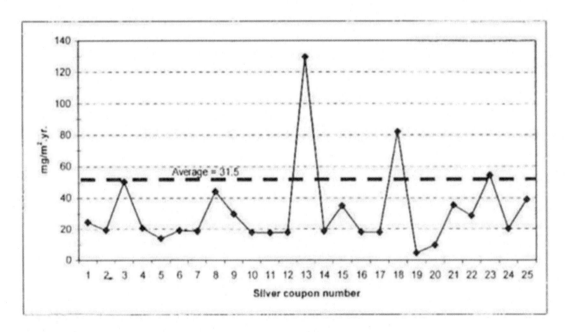

Figure 3. Weight-loss data for silver corrosion coupons at indoor conditions in a television manufacturing plant. Higher peaks correspond to the stations located at outdoor conditions.

Weight loss for indoor conditions ranged between 20 and 40 mg/m^2.yr, with an average value of 31.5 mg/m^2.yr. These results are in good agreement with the corrosion phenomena occurring on silver surfaces, where the formation of corrosion product films diminishes the corrosion rate regarding time. The corrosivity categories for silver, following ISO/CD 11844, shown in table 2, indicate that the most severe categories correspond to stations 4 (shipping area) and 5 (outdoor). In this case, hydrogen sulphide was detected in concentrations of 1.5 ppm at indoor and 2.0 ppm at outdoor conditions, but the H_2S concentration is not constant during the day. The presence of H_2S was detected during labour time, and it decreases after mid-evening to values undetectable by the sensor used (lower than 1 ppm). The average temperature in the plant was 22°C, and the relative humidity controlled in the range from 25 to 38 per cent. Records kept in the plant show that corrosion of silver increases during several periods when the temperature diminishes or the relative humidity increases. It is important to mention that the quality of the air supplied to the manufacturing process zone was not controlled. The indoor air supplied by several air conditioner handlers was insufficient to provide a positive pressure and to avoid the ingress of dust particles and other external environmental pollutants. The filters on the air system were for common use and incapable of retaining the chemical pollutants and also particulate matter in the range size from 2.5 to 10 micrometres.

Table 2. Corrosion categories for silver coupons exposed for one month. The last two categories correspond to the materials arriving zone and the outdoor stations, respectively.

Coupons	Corrosivity category	Visual observations
1, 2	IC1 to 1C2	Light gold—brown to metal blue
9, 10	IC1 to 1C2.	Light gold—brown to metal blue
17, 18	IC2 to IC3	Blue metal to dark brown
25, 26	IC3	Dark brown to grey
33, 34	IC3	Dark brown to grey

The SEM and EDX results show a repetitive corrosion behaviour, silver sulphide compounds predominate as corrosion products, and typical dust particles from 2 to 15 micrometres were detected on silver coupon surfaces, observed in figure 4. The morphology of a corroded silver surface and its chemical composition analysed by SEM and EDX are presented in figure 5.

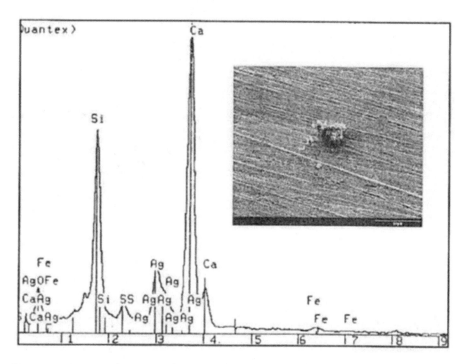

Figure 4. SEM and EDX analysis performed on a dust particle deposited on
the silver surface of a specimen exposed for one month at indoor conditions.
The EDX analysis corresponds only to the dust particle.

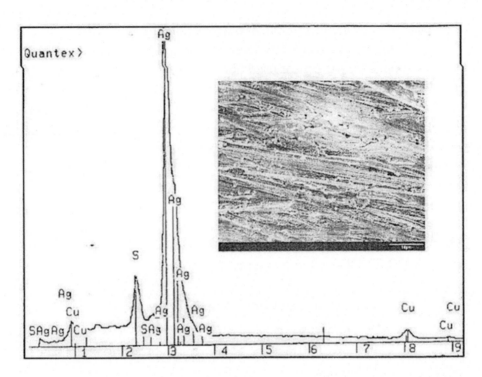

Figure 5. SEM. and EDX analysis performed on a corroded silver surface of a specimen exposed for one month at indoor conditions. EDX analysis corresponds to corrosion products.

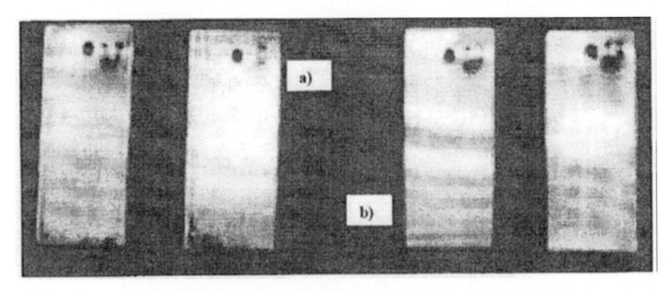

Figure 6: Silver specimens exposed to a ten-day corrosion test at indoor plant conditions: (a) without VpCI and (b) with VpCI.

The results for the application of VAPPRO 870 VCI are shown in figure 6. After ten days of testing, the silver specimens exposed without VCI under the indoor plant conditions exhibit tarnishing because of the formation of silver sulphide corrosion products, while the silver specimens protected with the VCI remain unaffected without corrosion. It is clear that the concentration of H_2S in the indoor plant conditions was enough to attack the silver surface. It is interesting to note that after twenty days of exposure in the laboratory environment, the silver specimens already tested in the absence of VAPPRO 870 VCI suffered from further corrosion, with increase in size of the black stains and the appearance of additional black stains, spread out on the specimen's surface. In the stamped numbers of the specimens, containing residual concentrated mechanical stresses, blue-green corrosion products become visible. This is a well-known phenomenon in specimens subjected to cold work, plastic deformation, or strong mechanical effects. On the other hand, the silver specimens that were tested in the presence of VAPPRO 870 VCI exhibit a clear, shiny surface, without stains even in the stamped numbers. This is due to the continued protective action of the thin adsorbed film of the corrosion inhibitor. The copper specimens remained without any surface changes after the ten-day assay, showing a very good corrosion resistance against the indoor environment, independently of the presence or absence of the VAPPRO 870 VCI.

A successful soldering test was performed using a rich tin lead-free solder and flux, attaching a copper wire to the non-corroded silver surface. When the soldering was applied to the corroded silver surface, it failed immediately.

The VCI was also tested in storage racks containing electronic components and located at different points of the TV manufacturing plant, with excellent results in the corrosion inhibition of silver and other metals.

9.10 Conclusions

Silver sulphide (Ag_2S) was the main corrosion product detected on silver specimens exposed at indoor plant conditions. Less than forty-eight hours were required to tarnish the silver surface by the action of sulphur compounds.

Silver sulphide severely affects the performance of soldering operations, avoiding a good electrical contact between silver pins and the frame board.

The sulphur compound responsible for the corrosion process was hydrogen sulphide (H_2S) produced by the exploitation of geothermal resources in the region.

The silver corrosion process was not influenced by the solutions and materials used during the fabrication of the TV boards. Also, no contamination caused by human activities was detected.

The absence of an air conditioning control system and adequate air filters were the main cause of contamination by particulate matter and gases in the indoor TV boards manufacturing zone.

The corrosion problem was resolved by the application of a vapour phase corrosion inhibitor (VCI) VAPPRO 870, supplied by Magna Chemical Canada Inc. The product prevented corrosion on silver components and retained their physical properties and the ability to solder these components.

9.11 References

[1]   J. D. Guttenplan. 'Corrosion in the electronic industry', *Corrosion*, ASM, 13, 1107 (1987).
[2]   'Corrosion costs and preventive strategies in the United States'. Supplement to *Materials Performance,* July 2002.
[3]   B. Miksic and A. Vignetti. 'Vapor corrosion inhibitors successful field application in electronics', *NACE Corrosion Conference,* 2000; Paper 00707.
[4]   C. Boit. 'The new millennium: challenges and opportunities', in *Microelectronics Failure Analysis,* R. J. Ross, C. Boit, and D. Staab (eds.), EDFAS, ASM, March 2001.

[5]   A. S. Brar and P. B. Narayan. 'Materials and processing failures in the electronic and computer industries', ASM, March 1998.

[6]   <u>A. Ortiz-Prado, R. Schouwenaars and S.M. Cerrud-Sanchez.</u>. 'Metodologia para la simulation acclerada del deterioro por corrosion atmosferica se presenta en equipo electronico', *Ingenieria, Investigation y Tecnologia,* 111(4), 145-156 (2002).

[7]   A. Ortiz-Prado, R. Schouwenaars, and S. M. Cerrud-Sanchez. 'Design, construction and tests of a system for accelerated simulation of damage by atmospheric corrosion on electronic equipment', *Corrosion Reviews* [this issue].

[8]   B. Valdez, N. Rosas, J. Sampedro, M. Quintero, J. Vivero, R. Velasquez, and G. Hernandez. 'Influence of elemental sulphur on corrosion of carbon steel in geothermal environments', *Corrosion Reviews,* 17 (3-4), 167-179 (1999).

[9]   ISO/CD 11844-1. 'Corrosion of metals and alloys—Classification of low corrosivity in indoor atmospheres—Part 1: Determination and estimation of indoor corrosivity', 2001.

[10]   H. Kim and J. Payer. 'Tarnish process of silver in 100 ppb 11_2S containing environments', *Journal of Corrosion Science and Engineering,* 1, Paper 14 (1999).

[11]   R. Baboian. 'Electronics corrosion', Chapter 70. in *Electronics.* Electrochemical and Corrosion Laboratory, Texas instruments Inc., 2001.

[12]   S. Zakipour, J. Tidblad, and C. Leygraf. 'Atmospheric corrosion effects of SO_2, NO_2 and O_3. A comparison of laboratory and field exposed nickel', *J. Electrochem. Soc.,* 144, 10 (1997).

[13]   T. Burleigh, Y. Gu, M. Vida, and D. Waldeck. 'Tarnish protection of silver using a hexadecanethiol self-assembled monolayer and descriptions of accelerated tarnish tests'. *Corrosion,* 57, 12 (2001).

GENERAL REFERENCES

•   R. P. Frankhental. 'Electronic materials, components and devices', *Ulig's Corrosion Handbooks,* Second Edition, R. Winston Revie (Ed.), John Wiley and Soons, 2000.

•   R. B. Comizzoli, R. P. Frankliental, and J. D. Sinclair. 'Corrosion engineering of electronic and photonic devices', in: *Corrosion and Environmental Degradation,* Vol. 19, Materials Science and Technology Series, Wiley-VCH, Weinheim, Germany <u>[year]</u>.

•   G. S. Frankel. 'Corrosion of microelectronic and magnetic storage devices', in: *Corrosion I'Vechanisms in Theory and Practice,* P. Marcus and J. Oudar (Eds.), Marcel Dekker Inc., New York, 1995.

CHAPTER

Proposal Case History on Preservation of Caisson Leg against Corrosion Caused by SRB

Nelson Cheng,[1] Benjamín Valdez Salas,[2] Dr Ernesto Beltran,[2] and Patrick Moe[1]

[1]Magna International Pte Ltd., 10H Enterprise Road, Singapore 629834
[2]Universidad Autónoma de Baja California (UABC), Mexicali, Baja California, Mexico

10.1 Abstract

This article describes the case history and background for the proposal for preservation of caisson leg against corrosion caused by sulphate-reducing bacteria (SRB).

Sacrificial anodes were used commercially to protect caisson legs against corrosion in seawater. However, there was a paradigm shift in 2000 after an explosion occurred in a caisson leg, which prompted an investigation to ascertain the cause.

Subsequently, NACE published Technical Paper No. 4200 under 'Protection of Offshore Platform Caisson Legs with a Vapor Corrosion Inhibitor—a Case Study' in 2014. The abstract of the paper states, 'In November 2000, gas build-up inside the confined compartment around an offshore platform caisson leg led to an explosion. The gas was found to be hydrogen generated by the depleted anodes inside the Caisson Leg. An investigation of the fatal explosion made several recommendations, including removal of the anodes, biocide-treated water and blasting grit that had accumulated inside the caisson legs over two decades.' Based on the above technical paper, the research team from Magna International and UABC has been approached by ADMA-OPCO to use vapour corrosion inhibitors to protect the caisson leg over a period of five years. The recommended proposal includes the use of VCI technology, VCI products, and a monitoring system to ensure the integrity of the recommended preservation methodology.

Keywords: caisson legs, VAPPRO VCIs, VAPPRO SRB-X, SRB, hydrogen sulphide

10.2 Findings from Magna International and UABC

Magna International and UABC have been tasked to consider said preservation with reference to Technical Paper No. 4200. While studying Technical Paper No. 4200, the research team discovered new findings and would like to highlight to all parties concerned that it is essential to include the protection of the caisson leg against SRB when submitting the technical proposal.

The research team from Magna International and UABC found out that the likely cause of the explosion in the caisson leg in 2000 was likely hydrogen sulphide rather than hydrogen gas as mentioned in Technical Paper No. 4200, because depleting of anodes will not result in significant presence of hydrogen gas unless the anodes were exposed to an acidic environment as shown in the chemical equation 2.

Anodic reactions involve oxidation of metal to its ions—for example, for anode, the following reaction occurs:

$$M \rightarrow M^{2+} + 2e, \qquad \text{(chemical equation 1)}$$

whereas M equals to sacrificial anode.

For cathodic reaction where anode is exposed in an acidic condition, where hydrogen ion is predominant, the following chemical reaction will take place as mentioned below and in figure A.

$$2H^+ + 2e \rightarrow H_2 \qquad \text{(chemical equation 2)}$$

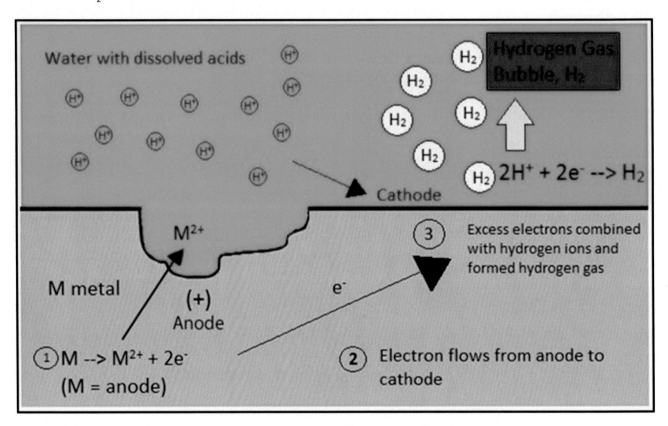

Figure A. Mechanism behind the formation of Hydrogen ions.

It is imperative to note that the lower explosive limit (LEL) and the upper explosive limit (UEL) of hydrogen are 18.3 per cent and 59 per cent, respectively.

Whereas the LEL and UEL of H_2S are 4.5 per cent and 45.5 per cent, respectively, making it more prone to explosion than hydrogen.

As seawater is alkaline and the hydrogen ion is not predominant, it is very unlikely that the depleting anodes will generate enough hydrogen gas that led to said explosion.

Sulphate occurs widely in seawater, which is a nutrient for sulphate-reducing bacteria (SRB); hence, seawater will promote the propagation of SRB and in turn will promote hydrogen sulphide (H_2S), as shown in the figure B mechanism below.

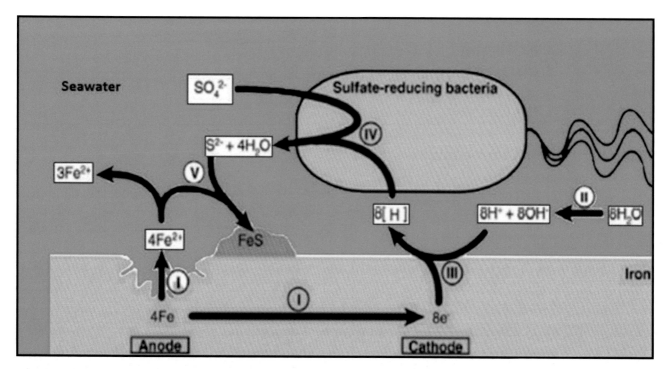

Figure B. Mechanism behind the formation of hydrogen sulphide, H_2S.

I. Dissociation of iron (Fe) because of anode and cathode arrangement
    $4Fe \rightarrow 4Fe^{2+} + 8e^-$

II. Dissociation of water in seawater
    $8H_2O \rightarrow 8H^+ + 8OH^-$

III. Electrons that transferred from anode to cathode combine with hydrogen ions and formed hydrogen atom.
    $8H^+ + 8e^- \rightarrow 8[H]$

IV. Sulphate-reducing bacteria (SRB) used the hydrogen atom in its respiratory process to reduce sulphate (SO_4^{2-}) in the seawater. Hydrogen sulphide is produced.
    $SO_4^{2-} + 4[H] \rightarrow H_2S + 2H_2O + 2OH^-$

V. The newly formed hydrogen sulphide reacted with iron ions (Fe^{2+}) and caused sulphuric corrosion (FeS).
    $Fe^{2+} + H_2S \rightarrow FeS + 2H^+$

Table 1. Common problems in caisson leg

Element	Description
High chloride content in seawater	Promotes high electrochemical activity in caisson leg, causing excessive flow of electrons from the anodic to cathodic points, resulting in severe corrosion
Presence of sulphate-reducing bacteria (SRB)	Promotes the presence of hydrogen sulphide (H_2S), resulting in sulphide stress cracking corrosion (SSC).

10.3 Corrosion in Marine Environment

Corrosion occurs when a substrate is in contact with an environment that produces a potential difference between the metal and its environment or within the metal itself. The metal then reacts with its environment—usually oxygen in air—to return to its natural occurring state, usually its oxides.

In the marine industry, the environment that causes corrosion is seawater. Seawater is a complex matrix that contains ions like sulphates, chlorides, calcium ions, etc., which are charged in nature (either positively or negatively). As such, it was reported that a solution full of charges promotes corrosion as it widens the potential difference between substrate and environment and within the substrate itself. It is especially notable that chlorides play the main role in the accelerating corrosion, as they are present as a majority as compared to other ionic species. Hydrogen sulphide is another cause of concern because of the reduction of sulphate presence in seawater by the sulphate-reducing bacteria.

As the marine industry seeks pre-emptive and preventive measures toward corrosion, the use of corrosion inhibitors—cathodic inhibitors such as sacrificial anodes in this case—is one of the methodologies.

10.4 Needs Statement

The inherent problem of H_2S in the caisson leg is a cause for concern and need to be arrested. VAPPRO 849 does not contain any biocides; as such, it does not contain any biocidal efficacy to combat sulphate-reducing bacteria (SRB).

The research team from Magna International and UABC would like to propose the use of VAPPRO SRB-X, which is based on glutaraldehyde $CH_2(CH_2CHO)_2$ chemistry in conjunction with the usage of VAPPRO 849.

VAPPRO SRB-X is the preferred biocide as its biocidal efficacy is not affected by the presence of H_2S, a paramount importance for combating SRB. In addition, it has lasting, lingering biocidal effect.

10.4 Objective

The objective of this proposal is to provide a comprehensible proven preservation methodology using pragmatic scientific monitoring system that will ensure the integrity of the preservation methodology over a period of five years.

10.5 Preservation Proposal

Taking into consideration the findings above and since VCI technology is the preferred technology by the client, the research team from Magna International and UABC would like to recommend the use of VAPPRO 849, VAPPRO VCI Pouch, and VAPPRO SRB-X for said preservation.

Propose the said preservation as follows:

I.   Empty caisson leg of seawater.
II.   By means of a pressurised spray gun, spray VAPPRO SRB-X into the caisson leg to remove remnant of sulphate-reducing bacteria.
III.   Sulphate-reducing bacteria (SRB) is anaerobic in nature; it will die when exposed to air. By means of an industrial air blower, introduce voluminous air into the caisson leg to speed up the drying process of VAPPRO SRB-X and enhance the killing rate of SRB.
IV.   Then fog VAPPRO 849 into the caisson leg by means of a blasting pot, just like fogging any powder. Please ensure no naked flame is allowed near caisson leg.
V.   The dosage for VAPPRO 849 is 10 grams per 28 litres; therefore, the required amount of VAPPRO 849 for caisson leg with the volume of 112,000 litres is equal to 10 grams x 112,000 litres divide by 28 litres = 40,000 grams, or 40 kilos of VAPPRO 849.
VI.   To reinforce the protection of the caisson leg over the period of five years, the use of VAPPRO pouch is recommended as it is a slow-release vapour corrosion inhibitor with lower vapour pressure than VAPPRO 849.
VII.   One VAPPRO pouch is effective to protect up to 1,410 litres of volume space; therefore, 112,000 litres of volume space in caisson leg requires 112,000 divided by 1,410, equal to 79.4 VAPPRO pouches.
VIII. About 80 VAPPRO pouches will be held in suspension uniformly by means of nets on the crossbar.
IX.   Please see figure C on how to install VAPPRO pouches in the caisson leg.
X.   Corrosion coupons are held in suspension uniformly on the crossbar to monitor the effectiveness of VCI products used and to measure the electrochemical activity of the preserved structures. Please see figure D on how to install corrosion coupons.

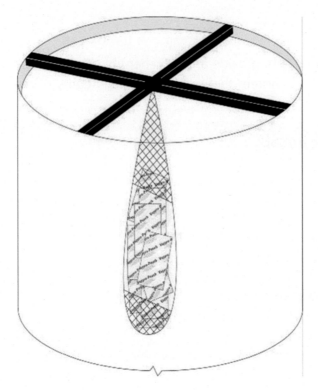

Figure C. Install VAPPRO pouches in the caisson leg.

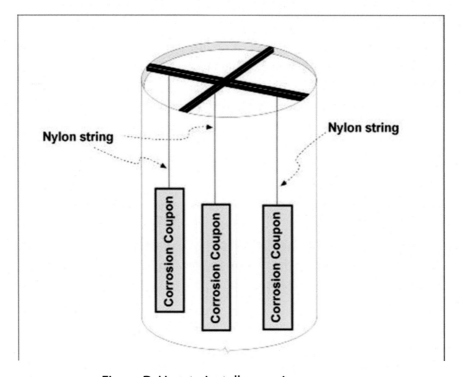

Figure D. How to install corrosion coupons.

10.7 Annual Corrosion Monitoring System of Corrosion Coupons

Corrosion rate is the speed at which any metal in a specific environment deteriorates. It also can be defined as the amount of corrosion loss per year mm in thickness (mmPy). The speed or rate of deterioration depends on the environmental conditions and the type and condition of the metal under reference.

It is found by R = d/t expressed in μm/y but can also be expressed in terms of weight loss g/m², mg/dm². day, oz/ft², among others. The total amount of lost thickness in micrometres is d = total. Loss occurrence is t = time in years.

Several pieces of data must be collected when calculating the corrosion rate of any given metal. Required data includes

- weight lost (the decrease in weight of the metal during the period of reference),
- density of the metal,
- total surface area initially present,
- length of time taken.

Corrosion rate is best expressed in terms of thickness or weight loss where the surface of the metal corrodes uniformly across the area that has been exposed.

This rate may vary if the rate expressed by the function above is used to compare corrosion rates for a period not less than one year with rates calculated over short periods. This is because the short time rates are prone to fluctuating environmental changes from season to season and from day to day. This method involves the exposure of a weighed piece of test metal or alloy to the specific environment for a specific time. This is followed by thorough cleaning to remove the corrosion products and then determining the weight of the lost metal because of corrosion.

The rate can also be calculated as follows:

R = KW/ (AT),
where
K = constant,
W = total weight lost,
T = time taken for the loss of metal.

10.8 Mechanism of VAPPRO 849 and VAPPRO Pouch

VAPPRO 849 is an anodic inhibitor that is a passivation inhibitor. It acts by a reducing anodic reaction, blocking the anode reaction and supporting the natural reaction of passivation metal surface, also forming an absorbed cohesive and insoluble layer on the metal.

The VAPPRO 849 anodic inhibitors reacts with metallic ions Me^+ produced on the anode, forming generally insoluble hydroxides, which are deposited on the metal surface as insoluble film and impermeable to metallic ions. From the hydrolysis of inhibitors results in OH^- ions, figure E shows the mechanism of the anodic inhibitor effect. It is very important that the VAPPRO 849 inhibitor concentration should be high enough as under-dosage of VAPPRO 849 affects the integrity of formation film protection.

In summary, the research team from Magna International and UABC wishes to bring to the attention of ADMA-OPCO that our technical preservation proposal is more comprehensive than the said recommendation mentioned in Technical Paper No. 4200.

Our proposal covers the usage of VCI technology and arrests the concern of sulphate-reducing bacteria, thereby eliminating the presence of H$_2$S and sulphide stress cracking corrosion (SSC) in the caisson leg.

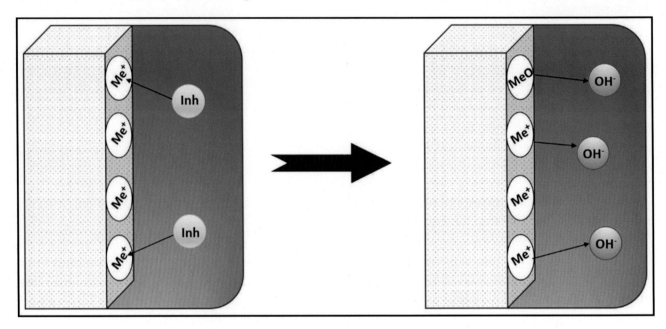

Figure E. Mechanism of the anodic inhibitor.

10.9 Sulphate-Reducing Bacteria and Hydrogen Sulphide Productions

Sulphate-reducing bacteria (SRB) are non-fermentative anaerobes that obtain their energy for growth from the oxidation of organic substances using inorganic sulphur oxy-acids or nitrate as terminal electron acceptors, whereby sulphate is reduced to sulphide. Biogenic sulphide may result in the corrosion of mild steel in an anaerobic environment.

For sulphate-reducing bacteria to produce hydrogen sulphide at such high concentrations, several conditions must be met. They include a sufficiently high sulphate concentration, the presence of sufficient biodegradable organic materials and nutrients, time, absence of oxygen (anaerobic), and reducing conditions for the microbial production of hydrogen sulphide gas.

The factors affecting SRB growth rates are as follows:

- temperature – general range: 20°C to 45°C. Optimum: 25°C to 38°C
- pH – general range: 6 to 9. Optimum: $7 \leq pH < 8$
- salinity requirements: SRB can survive and grow over a range of salinities. In addition, some SRB have a definite requirement for saline solutions. They are usually non-occurring in fresh waters.
- nutrients: In addition to the main metabolic requirement, many nutrients are also required. The common ones are phosphorus (dissolved phosphate ion), nitrogen (in the form of ammonia), calcium, and magnesium (found as dissolved salts).
- sulphide inhibition and adhesion of cells to the metal
- Limiting reactant is generally carbon source. However, in this case, sulphate ions could be the limiting reactant as well. It was found that the initiation of biocorrosion because of SRB only occurred in the presence of sulphate species tested on the corrosion of mild steel under different media both with and without sulphate ions. Additionally, it can be said that too high an initial

concentration of sulphate ions may inhibit the sulphate reduction rate of SRB, thus lowering the production rate of H_2S. H_2S, instead of hydrogen ion, could act as cathodic reactant—that is, $2H_2S + 2e^- \rightarrow 2HS^- + H_2$.

Adsorbed H_2 is removed to facilitate reduction of sulphate to sulphide by the bacteria hydrogenase. The activity of hydrogenase of SRB combines the adsorbed hydrogen atoms to produce H_2 gas first and then regenerates protons (H^+ ions). The produced sulphide would react with available proton to form H_2S.

10.10 VAPPRO SRB-X

Sulphate-reducing bacteria (SRB) are found in most oilfield systems, posing a serious challenge for effective control of microbial contamination in a production system. In view of the above, VAPPRO SRB-X was developed to combat said problem.

VAPPRO SRB-X is chemically stable in the presence of sulphides and organic acids, making it very effective to combat sulphate-reducing bacteria and acid-producing bacteria. It does not react with and is not deactivated by H_2S or other organic acids. The said chemical properties ensure that VAPPRO SRB-X, when added into the oilfield systems, will be fully available to act as biocide.

10.11 Some Common Applications of the Product Are Described Below

10.11.1 Water Flood Injection Water

VAPPRO SRB-X exhibits excellent stability in oilfield injection waters, which ensures that its antimicrobial activity will not be diminished in long pipelines. Hard waters and brines do not adversely affect its biocidal efficacy, and VAPPRO SRB-X is non-ionic, so it won't interfere with the action of demulsifiers, corrosion inhibitors, or surfactants. VAPPRO SRB-X is typically slug dosed into the injection water on a daily or weekly basis at 50 to 2500 ppm, active up to four hours, although the exact treatment regimen will depend on the condition of the system and the amount of water being treated.

10.11.2 Drilling, Completion, Work Over, and Fracturing Fluids

VAPPRO SRB-X functions as a biocide over a board pH range, and its efficacy is much faster at neutral to alkaline pH than at acidic ph. It is an excellent choice for use in preserving drilling muds and other oilfield fluids that are typically alkaline in pH. The combination of rapid alkaline efficacy at the typical use rate of 25 to 500 ppm as active and proven stability and effectiveness in high salinity matrices ensures microbial protection of these important fluids.

10.11.3 Produced Waters

Unlike some other biocides, VAPPRO SRB-X does not react and is not deactivated by H_2S or other organic acids. It is stable in the presence of sulphides and other organic acids commonly found in water flood injection system. VAPPRO SRB-X is typically added in slug doses on daily to weekly basis at a concentration of 50 to 2500 ppm as active.

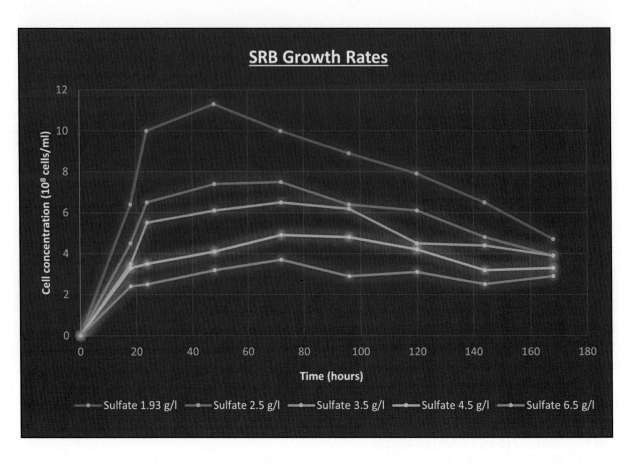

Figure F. SRB cell growth rates at different initial sulphate concentrations in the medium.

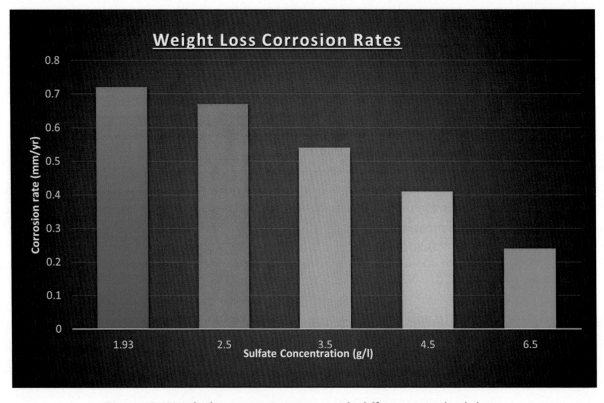

Figure G. Weight-loss corrosion rates with different initial sulphate concentrations in the medium at 37°C, one week after inoculation.

From figure F (above), sulphate concentration of 1.93 g/l peaked above the rest significantly after forty-eight hours, which translates to the highest corrosion rate (seen in figure F).

We can understand from figure F that SRB concentration was decreased as the initial sulphate concentration increases from 1.93 g/l to 6.5 g/l. Subsequently, as observed from figure G, a lower corrosion rate was obtained when the SRB growth was hindered.

Similarly, sulphate concentration of 6.5 g/l tends to inhibit SRB growths much better than the rest for the whole duration of the testing (figure F), hence giving rise to the lowest corrosion rates, as seen in figure G.

This behaviour can be attributed to the increasing toxicity of sulphates towards SRB metabolism or sulphate reduction.

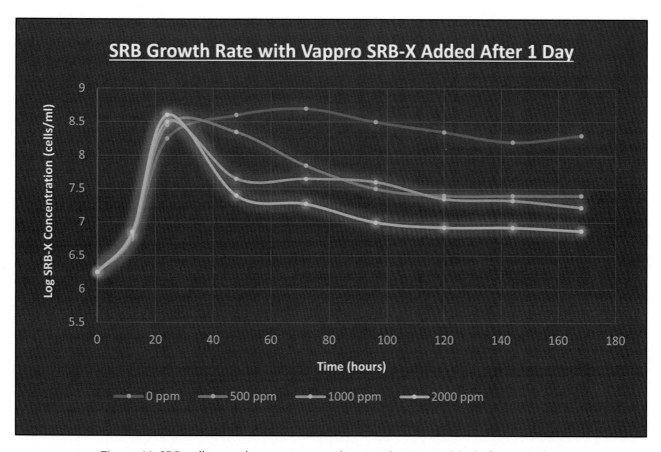

Figure H. SRB cell growth rates in a medium with SRB-X added after one day.

One day after inoculation, the culture was already found to have a good growth with an SRB cell count of 1.8×10^8 cells/ml. Therefore, a high VAPPRO SRB-X concentration was needed to achieve the inhibitory effect of SRB cell growth as seen in figure H.

10.12 References

1. Wen, J. Investigation of Microbiologically Influenced Corrosion (MIC) by Sulfate Reducing Bacteria (SRB) Biofilms and Its Mitigation Using Enhanced Biocides. in (2017).

2. Hu, A. Investigation of Sulfate-Reducing Bacteria Growth Behavior for the Mitigation of Microbiologically Influenced Corroslon (MIC). (Ohio University).

3. Guan, F., Zhai, X., Duan, J., Zhang, M., and Hou, B. Influence of Sulfate-Reducing Bacteria on the Corrosion Behavior of High Strength Steel EQ70 under Cathodic Polarization. *PLoS One* **11**, e0162315 (2016).

4. Telegdi, J., Shaban, A., and Trif, L. Microbiologically influenced corrosion (MIC). *Trends Oil Gas Corros. Res. Technol. Prod. Transm.* 191–214 (2017) doi:10.1016/B978-0-08-101105-8.00008-5.

5. Tran, T. T. T., Kannoorpatti, K., Padovan, A., and Thennadil, S. Sulphate-Reducing Bacteria's Response to Extreme pH Environments and the Effect of Their Activities on Microbial Corrosion. *Appl. Sci.* **11**, (2021).

6. Fletcher, L. E. Potential Explosive Hazards from Hydrogen Sulfide Production in Ship Ballast and Sewage Tanks. in (1998).

7. Paulus, W. *Directory of Microbicides for the Protection of Materials: A Handbook.* (Springer, 2005).

CHAPTER 11

The Process of Making VCI Paper, Using VAPPRO MBL 2200-Amino Carboxylate Corrosion Inhibitor (ACCI), and Ascertaining Its Impregnation Efficacy Dosage

Nelson Cheng,[1] Benjamín Valdez Salas,[2] and Patrick Moe[1]

[1]Magna International Pte Ltd., 10H Enterprise Road, Singapore 629834
[2]Universidad Autónoma de Baja California (UABC), Mexicali, Baja California, Mexico

11.1 Abstract

The purpose of this technical paper is to develop a process that can produce vapour corrosion inhibitor (VCI) paper that delays and prevents corrosion of ferrous metals. Kraft paper is impregnated with MBL 2200, a Magna International Private Limited product that contains 20 per cent amino-carboxylate corrosion inhibitor (ACCI). A humidity test was conducted on polished steel plates wrapped in kraft paper to test its corrosion inhibition properties. In this experiment, results were observed over five days. We concluded that 5 per cent amino-carboxylate corrosion inhibitor in kraft paper provides maximum corrosion protection compared to papers without corrosion inhibitor.

11.2 Introduction

Corrosion has been an anathema for various industries that rely heavily on metals. Storage and shipping after production are the main culprits that give rise to problems that stem from corrosion. Corrosion occurs on pure metal surfaces and alters the physical and chemical properties of the area it affected, rendering it useless.

Corrosion inhibitors are introduced to address this problem; by applying them to the metal's surface, the corrosion rate can be significantly reduced or even prevented. However, the effectiveness of corrosion inhibitors lies in the fact that there are no actual mediums to contain them. The corrosion inhibitor that was directly applied to the metals can escape to the surroundings easily, either because of the volatile characteristic like amines or it got accidentally scratched off while moving it.

Moreover, while shipping metals, the corrosion inhibitor may be accidentally scratched off. Thus, the VCI plastic film is invented. By impregnating VCI into plastic films, the VCI cannot escape easily. The plastic film isolates the metal from corrosive agents. The use of plastic, however, is not environmentally friendly. Although plastic can be recycled, not all of them go through recycling process. A portion of the plastic will be disposed because of the accident and immoral act of people; in the end, they wound up in the ground like the normal waste.

During their decomposition, they will release toxic and harmful substances to the soil, thus polluting the environment. Hence, this report postulates the replacement of plastic with paper and, by extension, aims to expound on a process of manufacturing VCI kraft paper.

11.3 Kraft Paper

Kraft paper is made up of chemical pulp that was produced from kraft process,[1] a sulphate process that converts wood into pulp. The process differs from normal paper production, for the solution that uses to convert wood into wood pulp consists of water, sodium hydroxide, and sodium sulphide. This chemical liquid mixture helps to break down the bonds between lignin, hemicellulose, and cellulose. As such, the amount of lignin present in the wood pulp is significantly reduced. As lignin weakens the formation of hydrogen bonds between the cellulose in the fibres when the paper is forming, the lower amount of lignin strengthens and increases the paper's resistance to wear and tear because of the successful formation of hydrogen bonds. However, because of the sulphate process, kraft paper contains some sulphur in it as residue.[2] The sulphur will react with moisture in the air and form hydrogen sulphide, which gives off a foul smell.

11.4 Corrosion Caused by Moisture

This corrosion is very common to take place if metals are exposed to environment without any protection. In this case, iron (Fe) will be taken as example because of its high usage in the industry. When iron is exposed to environment, it comes in contact with water moisture (H_2O) and oxygen. It then undergoes a redox reaction (rusting).[3] The rusting process first begins with the electrochemical process of iron donating ions to oxygen by reducing itself. The reactions are shown below:

$$Fe_{(s)} \rightarrow Fe^{2+}_{(aq)} + 2e^-,$$

$$O_{2(g)} + 4e^- + 2H_2O_{(aq)} \rightarrow 4OH^-_{(aq)}.$$

Iron ions (Fe^{2+}) then react with hydroxide ion (OH^-) to form iron (II) hydroxide ($Fe(OH)_2$):

$$Fe^{2+}_{(aq)} + 2OH^-_{(aq)} \rightarrow Fe(OH)_{2(s)}.$$

Last, rust ($2Fe_2O_3 \cdot H_2O$) is then quickly formed by the oxidation of iron (II) hydroxide:

$$4Fe(OH)_{2(s)} + O_{2(g)} \rightarrow 2Fe_2O_3 \cdot H_2O_{(s)} + 2H_2O_{(l)}.$$

11.5 Corrosion Caused by Sulphur

Kraft paper contains sulphur after its production process. The sulphur in the kraft paper will cause the rusting when it is converted to hydrogen sulphide upon contact with water moisture in the air.[4, 5] The products of the reaction will be hydrogen sulphide (H_2S) and sulphuric acid (H_2SO_4), as shown below:[5]

$$S_{8(s)} + 8H_2O_{(aq)} \rightarrow 6H_2S_{(g)} + 2H_2SO_{4(aq)}$$

Both of the products later on will react with metals and form rust. For the hydrogen sulphide (H_2S), it will adsorb onto iron and undergo certain reactions. The reactions are shown below:[6]

$$Fe_{(s)} + H_2S_{(g)} \rightarrow Fe_{(s)} + HS^-_{(aq)} + H^+_{(aq)}$$

There are two paths for this reaction according to Taylor's pairing:[6]

Path 1

$$Fe_{(s)} + HS^-_{(aq)} + H^+_{(aq)} \rightarrow FeHS^-_{(aq)} + H^+_{(aq)}$$

$$FeHS^-_{(aq)} + H^+_{(aq)} \rightarrow FeHS^+_{(aq)} + H_{(g)} + e^-$$

$$FeHS^+_{(aq)} + H_{(g)} + e^- \rightarrow FeS_{(s)} + H_{2\,(g)}$$

Path 2

$$Fe_{(s)} + HS^-_{(aq)} + H^+_{(aq)} \rightarrow Fe_{(s)} + S^-_{(aq)} + 2H^+_{(aq)}$$

$$Fe_{(s)} + S^-_{(aq)} + 2H^+_{(aq)} \rightarrow FeS_{(s)} + H_{2\,(g)}$$

The final product of FeS(s) is the rusting product.

As for the sulphuric acid, it will react with metal and form iron sulphate ($FeSO_4$), which is another rust product:[7]

$$Fe_{(s)} + H_2SO_{4(aq)} \rightarrow FeSO_{4(s)} + H_{2(g)}.$$

11.6 Vapour Corrosion Inhibitor

Corrosion inhibitor is a term introduced to the chemical substances that can delay or prevent corrosion of ferrous and non-ferrous metals. The corrosion inhibitors function to neutralise the corrosive agents or form a passivation layer on the surface of metals that prevent corrosive agents from reaching the metals.[8] A vapour corrosion inhibitor leans toward the latter. When it is put near or meets metals, it will react with the moisture of the surroundings and deposit a passivation layer on the metals that will prevent them from reacting with the corrosive agents.

Amino-carboxylate corrosion Inhibitor was chosen because it is less toxic compared to the commonly used corrosion inhibitors. It is also available in liquid form and is easier to apply compared to other corrosion inhibitors that are mostly in powder form.

The amino-carboxylate is also colourless in liquid form. Hence, it will not alter the appearance of paper or hinder the vision when viewing if it is put into transparent medium. Also, it does not taint the metals when it forms a protective layer on the surfaces, which will expedite routine inspection processes.[9]

11.7 Applying Method

There are several tests that are feasible to apply vapour corrosion inhibitor to the kraft paper. The processes are lithography[10] and impregnation. We illustrate the impregnation process in this article, as VAPPRO VCI paper is made via a common impregnation process.

11.8 Impregnation

There are two types of impregnations: roller impregnation and spray impregnation. The process of roller impregnation involves pulling papers from the paper mother roll, feeding it through an aerated bath of liquid vapour corrosion inhibitor before being squeezed by rollers and dried.

Both impregnation processes do not require precise amounts of liquid to be measured. However, impregnation process will have more wastage than lithography process, resulting in higher costs.

In this experiment, the roller impregnation method is adopted to impregnate kraft paper with the respective percentages of amino-carboxylate.

11.8.1 Roller Impregnation

An impregnation procedure is the most common and simplest method of loading porous kraft paper liquid VAPPRO MBL 2200, or more commonly wet impregnation.

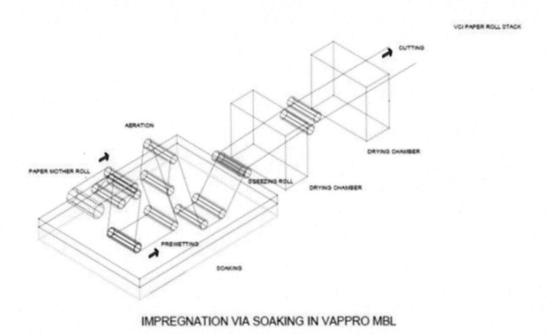

Figure 1. Impregnation via soaking in VAPPRO MBL 2200.

Figure 1 illustrates how mother-roll kraft paper is pre-wetted before it is soaked in the aeration bath of the VAPPRO MBL 2200, squeezed by two rollers before passing rapidly to the first dryer, and then to the second squeezed rollers before moving to the second dryer. Next, the impregnated VCI paper is cut into desired sizes or rolls.

The impregnation machine may consist of one section of squeezing rollers or one long drying tunnel, depending on the manufacturer.

In cases of short drying tunnels, the impregnation machine usually consists of two shot drying tunnels and two sections of squeezing rollers as shown in figure 2.

Figure 2. Spraying impregnation process.

11.8.2 Spraying Impregnation

Spraying impregnation involves pulling paper from the paper mother roll through a spray chamber where the kraft paper is impregnated via spraying. The paper is then passed through the drying chamber and allowed to dry before it is rolled into VCI paper rolls as shown in figure 3.

Figure 3. VCI paper production by spraying impregnation.

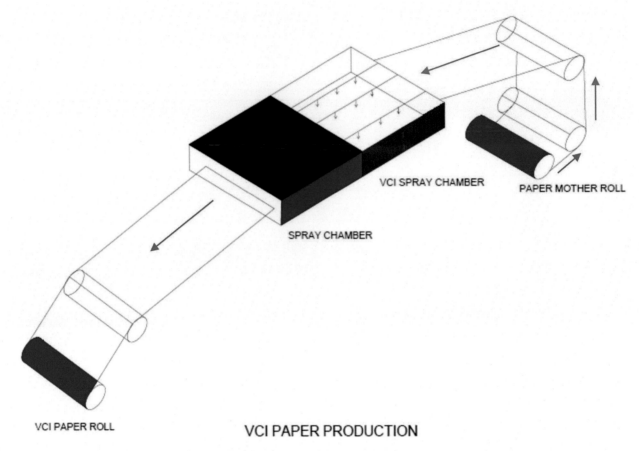

VCI SPRAY CHAMBER

SPRAY CHAMBER

PAPER MOTHER ROLL

VCI PAPER ROLL

VCI PAPER PRODUCTION

Figure 4. Spraying impregnation VCI paper production schematic diagram.

11.9 Determining the Effectiveness of MBL 2200

This experiment is to determine the effectiveness of MBL 2200 on ferrous steel when applied onto kraft paper. The experiment is carried out to determine the minimum amount of amino-carboxylate needed to have good protection on metals.

11.10 Experiment

Before carrying out the experiment to determine how many per cent of amino-carboxylate is needed to prevent corrosion, the amount of mixture between water and MBL 2200 is required.

11.11 Determine the Amount of Mixture

In this experiment, an A4-sized (210 mm x 297 mm) of MG Interleaving 57 gsm (grams per square meter) kraft paper was chosen to be the specimen. Because of lack of data on how much a kraft paper absorbs, 6 g of MBL 2200 mixture was decided to be put onto the kraft paper as a first attempt to find out if it is enough to cover the paper. The mixture was slowly put onto the surface by using a brush. After the first attempt, the area coverage by 6 g of mixture is only 65 per cent. Hence, by adding up to 100 per cent and considering wastage during brushing, 11 g of mixture is decided as the second attempt. After the second attempt, the 11 g mixture had completely covered the whole area of a new kraft paper.

Grams used for brushing (g)	Percentage coverage (%)
6	65
11	100

By using 11 g of the mixture,

$$\frac{\text{Grams needed for 1m}^2}{11\text{g}} = \frac{1000 \text{ x } 1000}{210 \text{ x } 297}$$

Grams needed for 1m² = 176 grams

= 0.176 kg.

Converting to grams per ft²,

1 ft² = 0.0929 m².

Grams needed for 1 ft² = 0.0929 x 0.176

= 0.0163 kg.

11.11.1 Materials and Apparatus Used

- 8 pieces of steel pieces with the dimension of 7.5 cm X 2 cm
- normal paper with the size of 12 cm x 10 cm
- 7 pieces of 12 cm x 10 cm interleaving kraft paper with 57 gsm
- MBL 2200, which is 20 per cent of amino-carboxylate corrosion inhibitor
- humidity chamber

11.11.2 Procedures

1. Eight steel pieces were polished with new no. 120 abrasive paper to remove all the grit and rust. Next, the metal pieces were polished with new no. 320 abrasive paper to leave a smooth surface. After that, the metal pieces were rinsed in ethanol to remove any organic stains (oil and grease) on the metal pieces. Then the metal pieces were dried.
2. Photos of steel pieces were taken.
3. Six pieces of each 12 cm x 10 cm kraft paper were impregnated with 2.1 grams of amino-carboxylate corrosion inhibitor with different percentage by roller impregnation method. For this testing, we used 1 per cent, 2 per cent, 3 per cent, 4 per cent, 5 per cent, and 20 per cent amino-carboxylate corrosion inhibitor of MBL 2200.
4. The steel pieces are wrapped individually, each with a different type of paper:
   - normal paper
   - kraft paper without amino-carboxylate corrosion inhibitor
   - kraft paper with 1 per cent amino-carboxylate corrosion inhibitor

- kraft paper with 2 per cent amino-carboxylate corrosion inhibitor
- kraft paper with 3 per cent amino-carboxylate corrosion inhibitor
- kraft paper with 4 per cent amino-carboxylate corrosion inhibitor
- kraft paper with 5 per cent amino-carboxylate corrosion inhibitor
- kraft paper with 20 per cent amino-carboxylate corrosion inhibitor, which is 100 per cent MBL 2200

5. Two steel pieces wrapped with normal paper and kraft paper without amino-carboxylate corrosion inhibitor are used as blank samples (control items). They have no amino-carboxylate corrosion inhibitor in them, and they are only used as a control/yardstick for the experiment.
6. The humidity chamber was turned on, and the temperature is set to 35°C.
7. When the temperature inside the chamber reached 35°C and humidity level reached 100 per cent, the specimens were placed in it.
8. The specimens were then taken out after one day, and the paper was carefully opened.
9. Any changes on the metals and papers were observed. Images were taken as a record.
10. The specimens were then wrapped again with the same paper and put back into the chamber.
11. Steps 8 and 10 were repeated for five days.

11.11.3 Observations

Type of paper used	Day 1 (before putting into humidity chamber)	Day 3	Day 5
Normal paper			
Kraft paper without amino-carboxylate corrosion inhibitor			

Type of paper used	Day 1 (before putting into humidity chamber)	Day 3	Day 5
Kraft paper with 1% amino-carboxylate corrosion inhibitor			
Kraft paper with 2% amino-carboxylate corrosion inhibitor			
Kraft paper with 3% amino-carboxylate corrosion inhibitor			
Kraft paper with 4% amino-carboxylate corrosion inhibitor			

Type of paper used	Day 1 (before putting into humidity chamber)	Day 3	Day 5
Kraft paper with 5% amino-carboxylate corrosion inhibitor			
Kraft paper with 20% amino-carboxylate corrosion inhibitor			

11.11.4 Table 1. Day 3 and day 5 corrosion rate

Type of paper used	% of corrosion on surface at day 3	% of corrosion on surface at day 5
Normal paper	15	50
Kraft paper without amino-carboxylate corrosion Inhibitor	40	60
Kraft paper with 1% amino-carboxylate corrosion inhibitor	1	1
Kraft paper with 2% amino-carboxylate corrosion inhibitor	1	1
Kraft paper with 3% amino-carboxylate corrosion inhibitor	1	2
Kraft paper with 4% amino-carboxylate corrosion inhibitor	0.5	1

Type of paper used	% of corrosion on surface at day 3	% of corrosion on surface at day 5
Kraft paper with 5% amino-carboxylate corrosion inhibitor	0.2	0.5
Kraft paper with 20% amino-carboxylate corrosion inhibitor	1	1

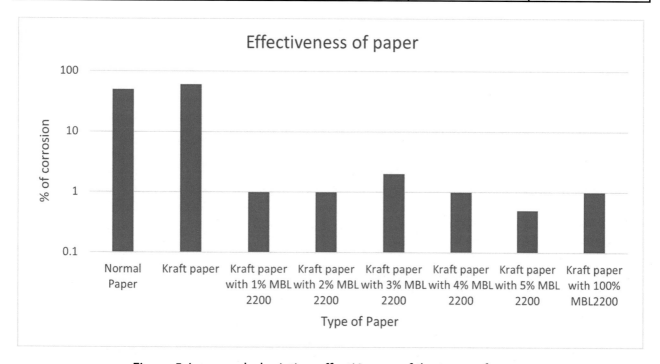

Figure 5. Log graph depicting effectiveness of the types of paper.

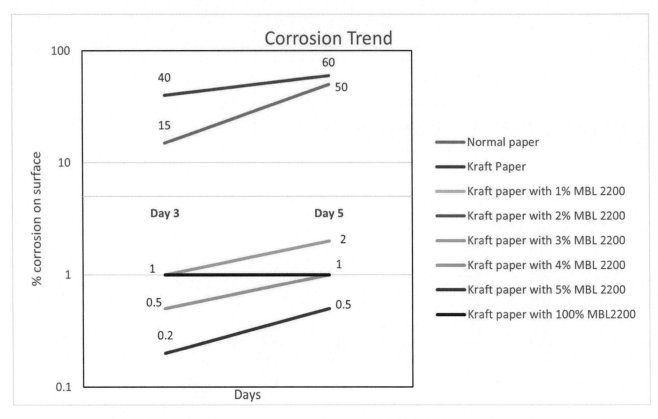

Figure 6. Log graph depicting corrosion trend and percentage comparison.

11.12 Conclusion

From the experimental results, we observed that 5 per cent amino-carboxylate has the better effect than 20 per cent amino-carboxylate. Hence, we concluded that kraft paper impregnated with 5 per cent amino-carboxylate has an excellent effect in protecting metals from corrosion observed after five days in wrapped carbon steel coupons exposed to extreme humidity conditions. Hence, the process of producing VCI kraft paper by impregnation in a liquid amine-carboxylate corrosion inhibitor solution method can be justified as success.

11.13 References

1. Review of New Source Performance Standards for Kraft Pulp Mills, EPA-450/3-83-017, U. S. Environmental Protection Agency, Research Triangle Park, NC, September 1983.
2. Minimizing the Sulphur Content in Kraft Lignin, Sara Svensson, ECTS 30.0, STFI-Packforsk, Stockholm.
3. Asato, R., date unknown. Corrosion, Internet Chemistry, Leeward Community College. Retrieved October 24, 2006.
4. The Role of Corrosive Sufur in Transformers and Transformer Oil, Lance R. Lewand, Doble Engineering Company, USA. Retrieved October 19, 2016, from http://www.doble.com/wordpress/wp-content/uploads/2002_3B.pdf.
5. Elemental Sulphur Corrosion of Mild Steel at High Concentration of Sodium Chloride, Haitao Fang, David Young and Srdjan Nešić, Institute for Corrosion & Multiphase Technology, Department of Chemical & Biomolecular Engineering, Ohio University, Athens.

6. Total Materia. (2014, June 26). Hydrogen Sulfide Corrosion. Retrieved October 19, 2016, from http://blog.totalmateria.com/hydrogen-sulfide-corrosion/.

7. BBC - Standard Grade Bitesize Chemistry - Reactivity of ... (2014). Retrieved October 21, 2016, from http://www.bbc.co.uk/bitesize /standard/chemistry /metals/ reactivity/ revision/4/.

8. Camila G. Dariva and Alexandre F. Galio (2014). Corrosion Inhibitors—Principles, Mechanisms and Applications, Developments in Corrosion Protection, Dr M. Aliofkhazraei (Ed.), InTech, DOI: 10.5772/57255. Available from: http://www.intechopen.com/books/developments-in-corrosion-protection/ corrosion-inhibitors-principles-mechanisms-and-applications.

9. Aminocarboxylate Salts as Corrosion Inhibitors in Coating Application, Kamlesh D. Gaglani, Cosan Chemical Corporation, Carlstadt, N. J., 24 March 1993. Retrieved October 19, 2016, from https://docs.google.com/viewer?url=patentimages.storage.googleapis.com/pdfs/US6127467.pdf.

10. Ryan (2002). Lithography. Retrieved October 19, 2016, from Technology Student, http://www.technologystudent.com/designpro/prtpro5.htm.

Combating Corrosion in Marine and Concrete Structures Synergistically Using Dual Protection— Colloidal Inhibition and VCI Adsorption

Nelson Cheng,[1] Benjamín Valdez Salas,[2] and Patrick Moe[1]

[1]Magna International Pte Ltd., 10H Enterprise Road, Singapore 629834
[2]Universidad Autónoma de Baja California (UABC), Mexicali, Baja California, Mexico

12.1 Abstract

The two most common causes of reinforcement corrosion are (i) localised breakdown of the passive film on the steel by chloride ions and (ii) general breakdown of passivity by neutralisation of the concrete, predominantly by reaction with atmospheric carbon dioxide. Sound concrete is an ideal environment for steel, but the increased use of de-icing salts in temperate countries, constant exposure of seawater, and the increased concentration of carbon dioxide in modern environments principally because of industrial pollution have resulted in corrosion of the rebar. The scale of this problem has reached alarming proportions in various parts of the world. The main purpose of this research was to establish the synergistically effect of the VAPPRO CRI 4600 corrosion inhibitor for seawater on various metallic materials: carbon steel, aluminium, and copper alloy at different concentrations via colloid formation and VCI adsorption using German VIA *TL 8135-002 Test* (Https://Wenku.Baidu.Com/View/218bb0cf0508763231121238.Html—Google Search, n.d.; Furman et al., 2004).

For the colloid formation, the changes in both physical and chemical properties of seawater, including pH, total hardness, alkalinity, total dissolved solids (TDS), and conductivity at different concentrations of VAPPRO CRI 4600, were assessed too. The test procedure involves dissolving the VAPPRO 4600 corrosion inhibitor (CI) powder in seawater to obtain a stock solution of 4 per cent VAPPRO CRI 4600 in seawater, which was further diluted to obtain the remaining concentrations. The analysis of parameters began when various metal species, polished beforehand, were placed into the solutions. The analysis was observed over a period of twenty-six days, and a total of nine sets of readings were obtained. From our observation near power stations burning fossil fuels, generating acidic rains, the pH can diminish to 6. In the case of the VCI adsorption test, a German TL 8135-002 was used. Adsorption of VAPPRO CRI 4600 organic molecules at the metal surface disrupts both the properties of the interface between the metal and its surrounding, which in turn effectively inhibits the corrosion process. By the adsorption of their ions via the hydrolysation of its molecules on the metal's surface, VAPPRO CRI 4600 adsorption inhibitors inhibit or reduce corrosion, resulting from polarisation of its ions on the metal. The effect of inhibitors adsorbed on metallic surfaces is to slow down the cathodic reaction as well as the anodic process of dissolution of the metal. Such an effect is obtained by forming a barrier to diffusion or by blocking reaction sites. Below are the details of the experiment test.

Corrosion of marine structures   **Corrosion in concrete structures**

Keywords: colloidal corrosion inhibitor, corrosion inhibition efficiency, VAPPRO CRI 4600 Volatile Corrosion Inhibitors, German TL 8135-002 Test, vapour inhibition test

12.2 Introduction

Corrosion in marine and concrete structures is a worldwide problem that strongly affects natural and industrial environments—in particular, the oil and gas industry. All its numerous and diverse facilities, equipment, and installations require products, methods, and techniques to protect, mitigate, and prevent corrosion damage Hummel 2014; Raichev et al. 2009). Corrosion inhibitors (CI) are one of the modern technologies applied for the management of corrosion for the benefit of the global economy (Garcia 2013).

The world and Mexico are undergoing an intense reformation process in the energy sector that is involving its oil, natural gas, and electricity industries. Abundant resources, such as deep-sea oil and shale gas, will be utilised, and additional refineries and pipelines will be built with the active participation of heavy foreign investments. The reform was recently approved by the Mexican parliament and is setting Petroleos Mexicanos (PEMEX), the national oil company, on the way to becoming a world oil enterprise (Layoza 2014).

12.3 Seawater Corrosion

The sea is a dynamic system in permanent motion, with complex surface currents and winds blowing over its surface, generating waves that reach the coast and its facilities and installations.

Seawater consists of a solution of many salts and numerous organic and inorganic particles in suspension. Its main characteristics are salinity and chlorinity and, from the corrosion point of view, dissolved oxygen (DO) content, which ranges from 4 to 8 mg/L, depending on temperature and depth. Its minor components include dissolved gases—CO_2, NH_3, and H_2S—found in seawater contaminated by urban sewage. The oceans house algae, bacteria, and phytoplankton, which generate about half of the oxygen in the atmosphere. Ocean surface salinity is determined by the balance between water lost by evaporation and water gained by precipitation. The salt concentration, particularly NaCl, varies from 2.0 per cent to 3.5 per cent, according to the sea location and the massive addition of fresh river water. For instance, in the Red Sea, an enclosed basin, salinity at high summer temperatures is 4.1 per cent; but in the Baltic Sea, it is about 2.0 per cent since many rivers feed into it.

Seawater is slightly alkaline, with a pH about 8; but when it is contaminated by acids, such as in coastal regions near power stations burning fossil fuels, generating acidic rains, the pH can diminish to 6.

12.4 Corrosion Inhibitors

In the last years, the use of CI is rapidly expanding worldwide for numerous technological and industrial applications: as cooling water systems (Schorr et al. 2012), steel-reinforced concrete, protected storage of military and electronic equipment (Valdez et al. 2003), acid pickling and cleaning (Carrillo et al. 2012), oil and gas industry, as additives to coatings, paints, and elastomers, for corrosion avoidance in oil pipelines (Hilleary et al. 2014; Murthy 2014).

The importance and relevance of the CI technology are evident by the many patents gathered in published reviews (Inzunza et al. 2013; Bastidas et al. 2005).

To prevent atmospheric corrosion, vehicles are covered during long periods with plastic sheets impregnated with vapour phase corrosion inhibitors (VPCI), also called volatile CI (VCI).

CI slows the rate of corrosion reactions when added in relatively small amounts to the treated system. They are classified into three groups:

- anodic inhibitors, which retard the anodic corrosion reactions by forming passive films
- cathodic inhibitors, which repress the corrosion reaction such as reduction of DO
- adsorption inhibitors, such as amines, oils, and waxes, which are adsorbed on the steel surface to form a thin protective film that prevents metal dissolution

12.5 VAPPRO CRI 4600, a Colloidal Corrosion Inhibitor

This polymolecular CI is added to seawater as a powder; then it converts into a colloidal suspension with nanoparticles dispersed in the water. These particles are adsorbed on the steel wall, forming a thin protective film. The performance of this inhibitor depends on physical, biological, and chemical factors.

The factors under analysis for this study include hardness, alkalinity, conductivity, and pH. Other factors such as dissolved oxygen etc. contribute as well but are not within our scope of this investigation.

We propose that the mechanism of colloidal formation functions by combining VAPPRO CRI 4600 with Ca^{2+} ions present in seawater to form an inert colloidal particle that is cationic in nature.

$$Ca^{2+} + \text{Vappro CRI 4600} \rightarrow Ca^{2+}\text{-Vappro CRI 4600 complex} \tag{1}$$

The colloidal particles formed adhere to the metal and prevent the onset of corrosion by preventing the loss of electrons. This causes the electrochemical cell to be incomplete and corrosion cannot occur. The CI VAPPRO CRI 4600 powder was specially developed to combat corrosion on mild steel and iron structures in stagnant seawater found in ballast tanks of ships and rigs.

In this study, the CI was tested to establish its effectiveness, the changes in both physical and chemical properties of seawater, which include pH, total hardness, alkalinity, and total dissolved solids/conductivity at different concentrations, with the purpose to find the optimum CI concentration and to provide recommendations on how the effectiveness of the inhibitor can be improved to reduce corrosion.

12.6 Results and Discussion

The practices recommended in the ASTM (ASTM 2013) and NACE (NACE 2000) standards were followed for evaluating the steel corrosion resistance. The measured weights for carbon steel show that at 0.05 per cent concentration, there was the least weight loss, indicating the least corrosion. Over the period of twenty-six days, the steel control specimen in seawater had lost 0.58 g, while those with inhibitor on the average had reduced the metal loss to about 0.10 g. This was even lower than the tap-water control of 0.15 g metal loss. The most effective CI concentration was 0.05 per cent as the metal loss was only 0.03 g (table 1).

Table 1. Effective CI concentration and the metal loss

Inhibitor concentration, %	Corrosion extent, g	Inhibitor efficiency, %
-	0.58	-
0.0125	0.19	22.6
0.025	0.11	81.0
0.05	0.03	94.8
0.10	0.05	91.3
0.25	0.09	84.4

The inhibition efficiency (IE) was determined using the equation

$$IE\% = \frac{M_u - M_i}{M_u} \, x \, 100$$

where M_u and M_i are the weight loss of the steel in uninhibited and in inhibited solutions.

12.7 Carbon Steel Corrosion Reactions

A drop-in water hardness was observed; however, this was not reflected in the conductivity. This means that other ions present in seawater had interacted with other than Ca^{2+} and Mg^{2+} ions. The proposed reactions include

VAPPRO CRI 4600 + Ca^{2+}/ Mg^{2+} → gelatinous white precipitate, (3)

VAPPRO CRI 4600 + Ca^{2+}/ Mg^{2+} + Fe^{2+} / Fe^{3+} → insoluble complex. (4)

As iron underwent the anodic reaction, the cathodic reaction expresses the oxygen reduction reaction.

$$Fe^{2+} \rightarrow Fe^{3+} + e^- \qquad\qquad\qquad \text{anodic reaction (5)}$$

$$O_2 + 4H^+ + 4e^- \rightarrow 2H_2O \qquad\qquad \text{cathodic reaction under acidic condition (6)}$$

$$O_2 + 2H_2O + 4e^- \rightarrow 4OH^- \qquad\qquad \text{cathodic reaction under neutral-alkaline condition (7)}$$

$$Ca^{2+} + HCO_3^- + OH^- \rightarrow H_2O + CaCO_3 \qquad\qquad (8)$$

$$Mg^{2+} + 2OH^- \rightarrow Mg\,(OH)_2 \qquad\qquad (9)$$

In all the above reactions, the reduction of the hydrogen ions or the production of hydroxyl ions raised the pH of the electrolyte in fresh water. Whereas, in seawater, the cathodic reduction observed by equations (8) and (9) produced a slightly alkaline surface condition that precipitated $CaCO_3$ and $Mg\,(OH)_2$.

In both mild steel pieces at 0.25 per cent and 0.1 per cent VAPPRO CRI 4600 in seawater, with pH range 5–6, dark pits were visually observed on the metal towards the end of the analysis. These pits were much more likely to be formed at the anodic area because of the formation of the precipitate layer.

At 0.025 per cent VAPPRO CRI 4600 in seawater and below, at pH 7.5–8.0, the steel pieces started to corrode. Thus, the inhibitor was not useful at such low concentrations.

At 0.05 per cent VAPPRO CRI 4600 in seawater, the pH range was about 7. With support from ferroxyl indicator test and weight loss test, it proved that there was optimum corrosion inhibition at this concentration of VAPPRO CRI 4600 in seawater, though with some staining of the metal (figure 1).

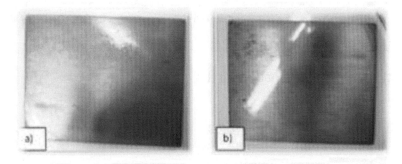

Figure 1. Carbon steel samples exposed to seawater and 0.05 per cent VAPPRO 844 before (a) and after (b) immersion in ferroxyl indicator.

12.8 Applications

Oil tankers, also called petroleum tankers, are ships specially designed and constructed at American and Asian shipyards for the bulk transportation of crude oil from the extraction sites (onshore and offshore) to the refineries. Smaller tankers are used to move refined products: fuel derivatives to the market. Since oil tankers were involved in damaging oil spills in the last years, they are strictly regulated and controlled (Heiderbach 2011; Chilingar 2008).

Petroleum steel tankers (figure 2) are cheaper and more efficient for oil transportation than submarine pipelines installed on the seabed—for instance, to deliver oil from North Africa to south Europe. On its way back, the tanker holds are full of seawater to provide adequate stability to the tanker (figure 3). CI is added to this ballast water. Pipes, storage tanks (figure 4), and pumps, using water for hydro test, are dosed with the same CI (Schorr et al. 2015).

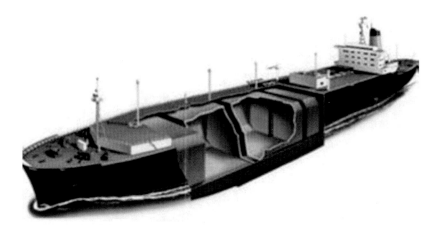

Figure 2. Petroleum transportation tanker showing holds.

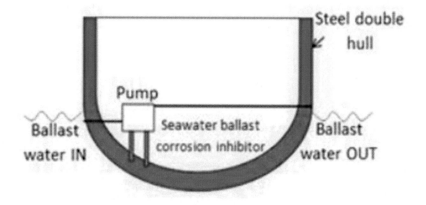

Figure 3. Ballast water tank.

Figure 4. Fire protection water storage tank.

12.9 Conclusion

- From the experimental observations, mild steel was protected with VAPPRO CRI 4600 corrosion inhibitors at 0.05 per cent inhibitor concentration, having only slight stains after a period of twenty-six days.
- Changes in seawater parameters were observed when the powder CI was introduced.
- VAPPRO CRI 4600 powder contributed to the increase of conductivity when it was introduced into the solution; however, when it reacted with the ions in seawater to form colloids, the conductivity dropped. The introduction of CI made the solution more acidic because of the mild acidic properties of VAPPRO CRI 4600.
- The higher concentration of inhibitor added the lower the alkalinity. For hardness, it showed that the calcium and magnesium ions were indeed used up in the reaction. This proved that the VAPPRO CRI 4600 powder followed the proposed reaction mechanism to form colloids.

12.10 VCI Adsorption Test of VAPPRO CRI 4600 Using German TL 81305-002

Controlling vapour pressure of VAPPRO CRI 4600 is the state of the art of VCI manufacturers, as different VCI carriers are used to meet each unique application because of the wide applications of VCI products. To use the principle of adsorption synergistically to prevent corrosion of metals, VAPPRO CRI 4600 should be evaluated for its ability to prevent vapour corrosion.

In view of the above, a reliable test method is essential. The German TL 8135-002 test method has been adopted to ascertain the vapour inhibition ability (VIA) properties of VAPPRO CRI 4600.

12.11 German TL 8135-003 Test (GmbH 2017)

A test sample with a high degree of sensitivity to corrosion through condensation water is packed together with a VCI auxiliary packing material in a vessel, which is then tightly closed. Condensation is then forced on the surface of the test sample. By means of a blank trial—that is, a trial structure of the same type but without VCI auxiliary packing material—it is determined whether the conditions of the trial are sufficient to cause corrosion to appear on the unprotected test sample.

12.11.1 Test Object

Four pieces of unalloyed solid construction steel

12.11.2 Test Sample

0.5 gram of VAPPRO CRI 4600

12.11.3 Test Solution

10 ml freshly prepared glycerine/water mixture with a density of 1.076 g/cm³ at (23±2)°C, which is intended to produce approximately 90 per cent relative humidity in the jar.

12.11.4 Test Equipment and Material

For each test, four test sets are necessary. A test set consists of the following parts:

(1) test jars, 1 L, wide-necked

(2) rubber stopper, 54 mm φ, with longitudinal through hole

(3) unalloyed solid construction steel test objects

(4) VAPPRO CRI 4600

(5) 10 ml freshly prepared glycerine/water mixture with a density of 1.076 g/cm³ at (23± 2)°C (glycerine/water mass ratio about 1:2)

(5) ethanol

12.11.5 Procedure of the Test

Four test objects were polished with 320 grit abrasive paper to remove all the grit and rust. Rinse with ethanol and dry them. The polished test object was inserted into the rubber stopper. Please see below figure A.

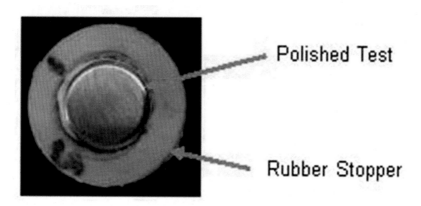

Figure A

Then rubber stopper was inserted to the test jar cover. A 0.5 gram of VAPPRO CRI 4600 was placed in the jar. Then the test jar was closed with jar cover. Please see below figure B.

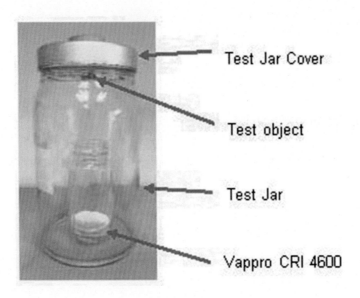

Figure B

For the blank sample, the test jar was sealed without inserting the VAPPRO CRI 4600. It had no VCI chemicals, and it is only used as a control/ yardstick for the experiment. Please see below figure C.

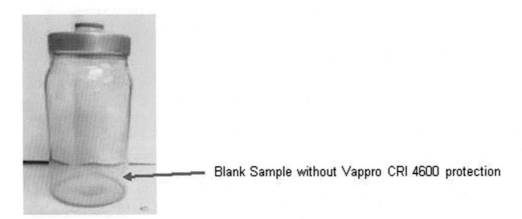

Figure C

The four test sets were stored for a period of (20±0.5) hours at a room temperature. At the end of the storage period, the jar covers were removed from the test jars; the freshly prepared 10 ml of the test solution, glycerine/water mixture, was poured into each jar immediately after opening; and the jars were immediately closed again. Please see below figure D.

After adding 10 ml of glycerin/water mixture

Figure D

After an additional 2 hours ± 10 minutes, the test jars were stored for a period of 2 hours ± 10 minutes in the heating chamber at temperature 40°C to create 90 per cent relative humidity in test jars.

On conclusion of storage in the heating chamber, the test objects were disassembled from test jars and dried with warm air. Then any signs of corrosion were inspected on the sanded surface of the test objects from the four jars.

12.11.6 Test Result

The sanded surface of the test object from the blank sample was badly rusted. Please see below figure E.

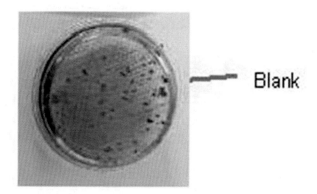

Blank

Figure E

No sign of corrosion was found on three test objects protected with VAPPRO CRI 4600. Please see below picture figure F.

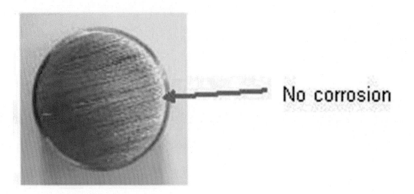

No corrosion

Figure F

Blank sample (without VAPPRO CRI 4600 powder)	Protected samples with VAPPRO CRI 4600		
Badly rusted	No visible corrosion was found on all three test objects.		

12.11.7 The Requirement of TL 8135-0002 for the Corrosion Protection Effect

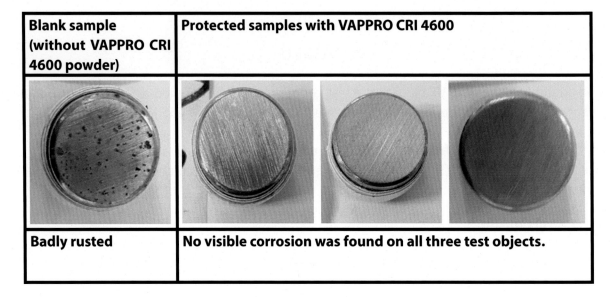

Evaluation of the test objects	Corrosion protection effect	
Keine korrosionsschützende Wirkung	**None**	(Grade 0)
Geringe korrosionsschützende Wirkung	**Slight**	(Grade 1)
Mittlere korrosionsschützende Wirkung	**Middle**	(Grade 2)
Gute korrosionsschützende Wirkung	**Good**	(Grade 3)

12.12 Conclusion

Based on above the test result, VAPPRO CRI 4600 passed the grade 3 German test method TL 8135-0002 and proven to have good vapour inhibition ability (VIA) properties for adsorption inhibition mechanism. In conclusion, it can be said that the dual mechanisms of colloid formation and VCI adsorption work synergistically together to protect marine and concrete structures from corrosion.

12.13 References

ASTM G31-13, 'Standard Practice for Laboratory Immersion Corrosion Testing of Metals' (West Conshohocken, PA: ASTM International).

Bastidas D. M., Cano E., Mora E. M., Volatile Corrosion Inhibitors: A Review, Anti-Corrosion Methods and Materials, Vol. 52, No. 7, pp. 71-77, 2005.

Carrillo I., Valdez B., Schorr M., Zlatev R., Inorganic inhibitors mixture for control of galvanic corrosion of metals cleaning processes in industry, NACE Conference, 2012, USA.

Chilingar G. V., et al., The Fundamentals of Corrosion and Scaling for Petroleum and Environmental Engineers (Houston, TX: Gulf Publishing Co., 2008).

Furman, A. Y., Kharshan, M., and Chandler, C. J. (2004). Performance and Testing of Vapor Phase Corrosion Inhibitors. Corrosion 2004, 1–12. https://www.onepetro.org/conference-paper/NACE-04418.

Garcia R., Valdez B., Schorr M., Eliezer A., Green Corrosion Inhibitor for Water Systems, MP 52, 6 (2013): pp.48–51.

GmbH, H. (2017). BFSV Verpackungsinstitut Hamburg GmbH Test Report Testing of corrosion protection effect of one VCI-film in accordance with TL 8135-0043 BFSV Verpackungsinstitut Hamburg GmbH. 49(2430), 43–46. https://www.controlox.eu/wp-content/uploads/2018/05/vci-single-wound.pdf.

Heidersbach R., Metallurgy and Corrosion Control in Oil and Gas Production (Hoboken, NJ: John Wiley and Sons, 2011.

Hilleary J., Dewitt J., Corrosion Rate Monitoring in Pipeline Casings, MP 53, 3(2014): p. 28.

https://wenku.baidu.com/view/218bb0cf050876323112123 8 html - Google Search. (n.d.). Retrieved December 8, 2022, from https://www.google.com/search?q= https%3A%2F%2Fwenku.baidu.com%2Fview%2F218bb0cf0508763231121238.html&oq=&aqs=chrome.0.35i39i362l8.371032055j0j15&sourceid=chrome&ie=UTF-8.

Hummel R., Alternative Futures for Corrosion and Degradation Research, Potomac Institute Press, 2014, pp. 2–13.

Inzunza R. G., Valdez B., Schorr M., Corrosion inhibitors patents for industrial applications. A review, Recent Patents on Corrosion Science 3, 2 (2013): pp. 71–78.

Layoza E., Building the PEMEX of tomorrow, Mexico oil and gas review, New Energy Connections, pp. 6–7, 2014.

Murthy T., Monitoring of Chemical Treatment Is Essential to Prevent Internal Corrosion, MP 53, 9(2014): p. 54.

NACE TM0169-2000, 'Laboratory Corrosion Testing of Metals' (Houston, TX: NACE International, 2012).

Raichev R., Veleva L., Valdez B., 'Corrosión de Metales y Degradación de Materiales,' Editor: M. Schorr, Universidad Autónoma de Baja California (2009), pp. 281–284. (Spanish).

Schorr M., Valdez B., Ocampo J., So A., Eliezer A., Materials and corrosion control in desalination plants, MP 51, 5 (2012): pp. 56–61.

Schorr M., Valdez B., Salinas R., Ramos R., Nedev N., Curiel M., Corrosion Control in Military Assets, International Material Research Congress 2015, Symposium 6J NACE: Corrosion and Metallurgy, Cancun, Mexico.

Valdez B., Flores J., Cheng J., Schorr M., Veleva L., Application of vapour phase corrosion inhibitors for silver corrosion control in the electronic industry, Corrosion Reviews 21, 5-6 (2003): pp. 445–457.

13

Food Industry: Equipment, Materials, and Corrosion

B. Valdez,[1] M. Schorr,[1] R. Salinas,[1] and A. Eliezer[2]

[1]Universidad Autónoma de Baja California (UABC), Mexicali, Baja California, Mexico
[2]Beer Sami Shamoon College of Engineering—Sheva, Israel

The food industry's production plants, equipment, and materials are affected by corrosion because of their damp environments. Corrosion control is achieved by the selection of corrosion-resistant materials and application of sanitising procedures.

The three largest markets worldwide, according to the extent of their production, number of consumers, and economic and social significance, are the food, energy, and water industries. The food market is the most important and the largest, which includes all the inhabitants of this planet—about eight billion—because everyone eats.[1-2]

Techniques for preserving food from biodeterioration following harvest or slaughter, including drying, salting, fermenting, and pickling, have been known since prehistoric times. Modern techniques include canning, freezing, dehydrating, cooking under vacuum, and adding chemical preservatives.

Today, many cleaning and sanitising agents are employed to remove bacteria, scale, fouling, and biological and mineral deposits. This great variety of corrosive environments and aggressive chemical agents requires the use of corrosion-resistant alloys. They are widely applied in the numerous sectors of this industry to avoid and/or minimise corrosion.[3-4]

13.1 Food Processing Corrosion

The physicochemical characteristics of the raw materials and processed foods have different corrosivity levels that depend on their chemical composition, texture, and interaction with the processing, packaging, and transportation equipment (e.g., trucks constructed of white painted steel, Al-Mg Grade 5652 alloy [UNS A95652], etc.).

Foods are classified into three groups according to their corrosivity:

- noncorrosive: cereals, oils, fats, meats, fish, and milk
- mild corrosivity (foods with a pH of 6 to 7 and <1 per cent of sodium chloride [NaCl]): unfermented dairy products, fruit syrups, wines, beers, soups, canned meats, and sweet carbonated drinks
- high corrosivity (foods with a pH of 3 to 5): citric fruit juices, ham, acidic canned fruits, hot gravies, sauces and dressings, and vegetables and fish pickled in brines containing 1 to 3 per cent salt or diluted vinegar (acetic acid [CH_3COOH])

Table 1 shows the pH ranges for different foodstuffs pertaining to important sectors of the food industry. Many liquid foods are consumed as aqueous solutions, and they contain chloride salts that could be concentrated by evaporation during processing, which would induce pitting corrosion on localised areas of metal surfaces.

13.2 Food Taste and Corrosion

Food consists mainly of proteins, carbohydrates, and fats. Processed foods contain many additives to improve their appearance, quality, and preservation. They have a wide pH range (table 1) as well as varying water, salt, and acid (principally vinegar) content that affects their corrosivity. Foods have many tastes that are based on combinations of sweet, sour, salty, bitter, spicy, and astringent. From the corrosion point of view, however, the most relevant are salty, sour, and sweet.

Table 1. Ranges of pH values for different foodstuffs

Foodstuff	pH range
Vegetables	3.0 to 6.0
Fruits	2.0 to 5.0
Bakery	5.0 to 6.5
Meats	6.0 to 7.0
Fish	5.5 to 6.0
Dairy	5.0 to 6.5
Beverages	2.0 to 5.5

13.2.1 Water

Vegetables and ripe fruit contain between 70 and 90 per cent water. Water is a medium for preparing, cooking, and baking foods such as confections and hot and cold beverages. Considered the universal solvent, water can dissolve acids, bases, salts, sugars, and some alcohols to form bonds such as hydronium ions (H_2O^+), which are involved in corrosion of metallic cooking and container equipment. Water's interaction with food depends on the food's physicochemical properties, and its molecular and ionised structure affects corrosion.

13.2.2 Salt

Several branches of the food industry, such as meat packaging, sausage making, fish curing, and food preservation, employ NaCl as a preservative, seasoning, or both in the same vessel. Up until the nineteenth century, when industrial food freezing and canning were introduced, commercial and navy ships preserved meat in wooden barrels filled with granular salt for their long voyages since putrefaction bacteria cannot live in salt. Salt is a hygroscopic substance that absorbs atmospheric moisture and forms a concentrated salt solution or slurry, which can corrode steel and deteriorate industrial food processing plant floors made of concrete or ceramic tiles.

13.2.3 Acids

Acids in water solutions impart a sharp, sour taste to food and have a corrosive action on steel. Therefore, food containers should be fabricated from mild steel that is protected with corrosion-resistant coatings, stainless steel (SS), or plastic materials. Most of the vegetables and fruits consumed by humans are acidic (table 1). Three acids: acetic acid (vinegar), citric acid ($C_6H_8O_7$), and phosphoric acid (H_3PO_4) are widely applied in the food industry for preserving and acidifying beverages.[5]

13.2.4 Sweet

The sweet taste is imparted by various sugars obtained from sugarcane, beets, dates, honey, and maple syrup. Sugar is a prime source of energy in the human diet and is needed to maintain body temperature and activity. A solid or concentrated syrup is used to manufacture sweet foods such as confections, candies, and all types of juices. Sugar dissolves easily in water to form a molecular solution with minimal electrical conductivity, which is only slightly corrosive because it contains some quantity of dissolved oxygen.

Table 2. Stainless steel used in the food industry					
UNS[(A)]	Chemical composition %w/w				Characteristics/Uses
	Cr	Ni	Mo	Cmax	
Martensitic and ferritic chromium steels					
S41600	12-14	-	-	0.15	Easily machinable/valve stems, plugs, and gates
S42000	12-14			0.15	Hardenable by heat treatment/cutlery, and cladding over steel
S43000	16-18			0.2	Good corrosion resistance/structural purposes
Austenitic chromium-nickel steels					
S30200	17-19	8			Good corrosion resistance/general purposes
S30400	18-20	8-12		0.08	Good corrosion resistance/dairy equipment
S31600	16-18	10-14	2-3	0.10	Superior corrosion resistance/the workhorse of the dairy industry
[(A)]UNS: Unified Numbering System					

13.3 Food Industry Sectors

The food industry consists of numerous sectors based on the varied food resources—from animal, vegetable, and mineral. This article, however, concentrates on the dairy, beverage, and canned food sectors.

13.3.1 Dairy

The dairy industry collects, treats, and distributes milk used as feedstock for manufacturing cheese, butter, yogurt, and cream. Corrosion in these plants depends on the nature of the products and the processes involved. Milk is an emulsion/suspension mixture of lactose, proteins, fat globules, minerals, and vitamins. Bacterial fermentation processes convert lactose into lactic acid during souring.[6]

Milk is typically pH neutral; it is the lactic acid that is responsible for most corrosion attack. The dairy equipment and accessories are constructed from smooth, corrosion-resistant materials that are easily cleaned. SS (table 2) is the material selected so the odour, flavour, and colour of milk products are not affected. A list of the main SS equipment installed in a modern dairy plant is shown in table 3.

Table 3. Typical stainless steel dairy equipment	
Milking machines	Homogenisers
Vacuum pumps	Pasteurisers
Centrifugal pumps	Tubular and plate heat exchangers
Agitators and mixers	Vacuum evaporators
Milk coolers	Milk dryers
Bulk milk storage tanks	Spray milk dryers
Clarifiers	Conveyors
Cream separators	Vessels for cleaning solutions
Centrifuges	Piping and tubing equipment fittings
Bulk milk tankers	Cleaning-in-place spray devices

SS is resistant to cooling waters, which are used in food processes at various pH values.' Most containers, pipework, and food processing equipment are manufactured from Type 304 (UNS S30400) or Type 316 (UNS S31600) SS alloys. Ferritic SS alloy with 17 per cent chromium is used widely for splash backs, flexible hoses, and equipment, where corrosion resistance properties are less in demand.

13.3.2 Beverages

Beverages include beer and wine made by fermentation of barley or grapes, respectively; alcoholic spirits produced by distillation after fermentation of grains and fruits; and soft drinks that are carbonated and non-carbonated.[8] These beverages are mainly supplied in coloured transparent glass jars and bottles (figure 1) and in aluminium or steel-plated cans that are sometimes lined with plastics. Glass, for all types of food containers, is considered the most chemically inert packaging material. Nowadays, wine bottling machinery and wine fermentation tanks are constructed from SS, mainly UNS S31600 (figure 2).

Nanotechnology is applied to produce and stabilise nanoparticles used in the food and related industries. Beverages and fruit juices, which may lose some properties during pasteurisation processes performed at elevated temperatures, are protected by the addition of nanoparticles to maintain their vitamins and minerals.[9] The production of beverages involves the use of great amounts of water in the cleaning, storing, bottling, and canning procedures. These wet, damp, and highly humid conditions contribute to plant corrosion and premature equipment failure. The use of SS helps prevent the occurrence of these noxious corrosion events.

Pure food-grade H_3PO_4 is an additive for sauces, mayonnaise, and fruit juices. It is used to acidulate cola-type beverages, impart a slightly sour taste, and prevent sedimentation of solid iron oxides. It is also added to soft cheese as a calcium phosphate to avoid water separation. The processed fruits and vegetables produced for long-time preservation are hermetically sealed, sterilised by heat, and stored in glass jars or mainly supplied in coloured transparent steel cans.

13.3.3 Canned Food

In the United States, the Institute of Food Technologists (Chicago, Illinois, USA), the Can Manufacturers Institute (Washington, DC, USA), and the Steel Recycling Institute (Washington, DC, USA) support the canning industry, which includes producers, sellers, and distributors of cans made from tinplate, black steel plate, and aluminium.

Seafood, such as sardines, tuna, salmon, mackerel, and oysters, is packed in oil or tomato sauce in different forms and sizes of tin- and plastic-coated steel cans. The sardine canning industry, which started on the US central California coast in Monterey, is supported by the cool Pacific waters that are rich in nutrients.[10]

The general trend of the influence of acidity on corrosion of tinplate is shown in figure 3. Acidic corrosion attacks the internal surfaces of the cans. As pH increases and acidity decreases, the corrosion rate is significantly reduced. Filiform corrosion frequently appears in steel cans and is characterised by a filament pattern with several source points. This process is promoted by the osmotic action of water from the atmosphere. An active filament is a line of corrosion products, caused by corroding steel substrate under the coating film, that travels in random directions.[11]

Figure 1. SS wine bottling machinery.

Figure 2. SS fermentation tanks.

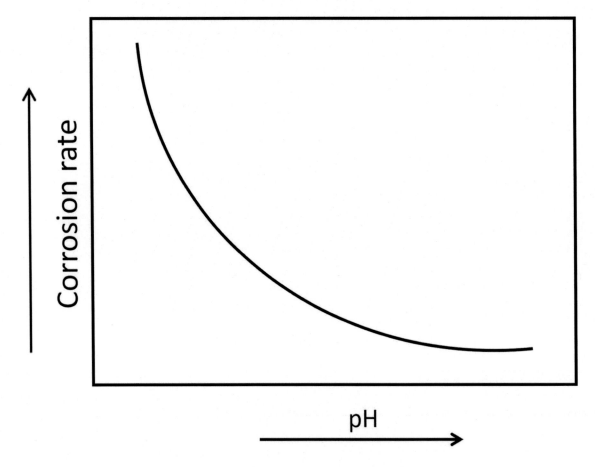

Figure 3. Acidity effects on corrosion of tinplate cans.

13.4 Cleaning and Sanitising

Hygiene and sanitation are basic requirements of the food industry. Foodstuffs contain nutrients that provide an ideal medium for bacterial growth, which can promote microbiologically influenced corrosion; corrosion under acidic, salty, fatty, and fouling deposits; and calcareous scaling from cooling water on heat transfer surfaces.

Cleaning is a mechanochemical operation for disinfecting food processing machinery so product contamination as well as damage to the SS surfaces are prevented. SS resists corrosive attack from chemicals formulated for cleaning and sanitising, such as chlorine, alkalis, mixed acids, organic ammonium quaternaries, halogenated hydrocarbons, and detergents.[12]

Several institutions provide data, standards, and guidance on sanitising procedures, including NSF International (Ann Arbor, Michigan, USA), International Association of Food Industry Suppliers (Santa Fe, New Mexico, USA), International Association for Food Protection (Des Moines, Iowa, USA), and NACE International (Houston, Texas, USA).

13.5 Conclusions

Traditional festivities of ancient and modern nations are based on the consumption of principal foods—grains, meats, vegetables, fruits, and beverages—but with the addition of herbs and spices to improve their taste.[13] The corrosion implications associated with these foodstuffs, along with sanitising procedures applied to equipment to ensure their cleanliness, efficiency, and profitability, must be considered for equipment reliability.

Corrosion prevention and control require the application of appropriate methods and techniques from the early stages of design through the construction, erection, and operation of food processing and production plants. Corrosion resistance is the main property to be considered in the choice of construction materials for such plants, but the final selection must be a compromise between technological and economic factors.

13.6 References

1. B. Valdez, ed., Food Industrial Process, Methods and Equipment (London, U.K.: InTechOpen, 2012).
2. B. Valdez, et al., 'Corrosion Control in Industry:' Environmental and Industrial Corrosion,' B. Valdez, M. Schorr, eds. (London, U.K.: InTechOpen, 2012).
3. R. Raichev, L. Veleva, B. Valdez, 'Corrosion de Metales y Degradacion de Materiales,' M. Schorr, ed. (Mexicali, Mexico: Universidad Autonorna dc Baja California, 2009).
4. P. R. Roberge, Handbook of Corrosion Engineering, Chapter 8: 'Materials Selection,' (New York, NY: McGraw-Hill, 2008).
5. M. Schorr, B. Valdez, 'The Phosphoric Acid Industry: Equipment, Materials and Corrosion,' Corros. Rev. 34, 1-2 (2015): pp. 85–102.
6. M. Schorr, D. Puga, 'Stainless for Corrosion Control in the Dairy Industry' Stainless Steel World, October (2000).
7. B. Valdez, M. Schorr, 'Stainless Steels for Corrosion Control in the Food Processing Industry,' Stainless Steel World (2004), pp. 419–422.
8. B. Valdez, et al., 'Wineries: Equipment, Materials and Corrosion,' MP 54, 4 (2015): pp. 68–71.

9. P. Sanguansri, M. A. Agustin, 'Nanoscale Materials Development—ii Food Industry Perspective,' Trends in Science & Technology 17, 10 (2006): pp. 547–556.

10. R. Gore, 'Between Monterey Tides,' National Geographic Magazine 177, 2 (1990).

11. G. Lopez, B. Valdez, M. Schorr, 'Micro and Nano Corrosion in Steel Cans Used in the Seafood Industry,' Scientific, Health and Social Aspects of the Industry B. Valdez, M. Schorr, R. Zlatev, eds. (London, U.K.: InTech, 2011).

12. A. Farook, 'Peculiar Corrosion Behavior of Type 316L in Simulated Cooling Water Various pH Values,' MP 53,10 (2014): pp. 44.

13. M. F. Vamosh, Food at the Time of the Bible. From Adamk tipple to the Last Supper (Herzliya, Israel: Palphot Ltd., 2007).

CHAPTER 14

Green Corrosion Inhibitors for Water Systems

B. Valdez,[1] M. Schorr,[1] R. Garcia,[1] and A. Eliezer[2]

[1]Universidad Autónoma de Baja California (UABC), Mexicali, Baja California, Mexico
[2]Beer Sami Shamoon College of Engineering—Sheva, Israel

14.1 Abstract

Corrosion affects the durability of civil infrastructure assets, including water production, supply, and storage systems. 'Green' corrosion inhibitors will extend the life of industrial equipment for water systems. Special green inhibitors are obtained from plants growing in desert regions of the state of Baja California, Mexico, by ethanolic and aqueous extraction.

Environmental quality, worldwide water scarcity, and clean energy have been established today as central disciplines in modern science, engineering, and technology. They are already being linked to the critical problems of climate change, global warming, and greenhouse gas emissions, all interrelated phenomena.[1-2] Furthermore, it is now generally accepted that corrosion and pollution are harmful processes that are interrelated, since many pollutants accelerate corrosion, and corrosion products such as rust also pollute bodies of water. Both are pernicious processes that impair the quality of the environment, the efficiency of industry, and the durability of the water infrastructure. In this time of energy crisis and economic turmoil, it is essential to develop and apply safe, 'green' (environmentally friendly) corrosion inhibitors.[3]

14.2 Water Systems

Fresh water comes from rain and snow; it accumulates in rivers and lakes and generally contains <1,000 mg of dissolved solids per litre (mg/L). Potable and building water include low levels of total dissolved solids (TDS), and some chemicals (e.g., chlorine) are added for health reasons.

Many types of water are produced, transported, and used: potable for municipal systems, irrigation for agricultural crops, and cooling for industrial facilities, facilities using fossil fuels, and nuclear energy. Figure 1 depicts a plant for treatment and clarification of water for human consumption.

Figure 1. A plant for water treatment and clarification.

Water is conveyed by a pipeline system, which consists of a large number of pipes, pump stations, and valves, that moves the water from a source to the consumption location. Pumps are essential components of water supply systems (figure 2). Usually, water pipelines are fabricated from ductile iron (DI) and from carbon steel (CS) based on the American Petroleum Institute (API) standards, but they may also be constructed from concrete or plastics, including reinforced plastics.[4] A system of painted steel water pipelines is shown in figure 3.

Figure 2. Pumps for water transportation.

Figure 3. Painted steel water pipelines.

Water pipes have an inner diameter between 0.10 and 2.0 m, and the water flows at speeds of 1 to 6 m/s. Modern water pipelines are operated remotely from computerised control rooms, and satellite surveillance is used to detect leaks or mechanical failures. The water quality and its influence on human health depends on the pipeline performance and whether it is free from corrosion, scaling, and fouling.

Steel corrosion is an electrochemical process that occurs on a pipe surface upon reaction with the water components, mainly dissolved oxygen (DO) and salts. Waters with a high concentration of dissolved and suspended solids, such as carbonates, silicates, phosphates, and hydroxides, form thick scales that might plug the pipes and interfere with water flow. Sometimes, macro and microorganisms thrive in water and cause corrosion.

The water corrosivity is determined by laboratory corrosion tests, simulating industrial conditions, and applying ASTM standards 5 and NACE TM0169.[6]

14.3 Corrosion Protection and Control

The water infrastructure requires the application of corrosion control methods and techniques from the early stages of design through the construction and operation of the equipment. Practical methods that minimise or eliminate corrosion include the selection of corrosion-resistant construction materials, application of coatings and linings, cathodic protection (CP), and use of corrosion inhibitors. The most direct means of preventing corrosion is the choice of suitable materials. The final selection, particularly for water pumps, must be a compromise between technological and economic factors.[7]

The purpose of a coating or lining is to act as a non-reactive barrier between the water and the material to be protected, generally steel (figure 4). Coatings fall into three main groups based on their chemical nature:

metallic, organic (including paints), and inorganic. CP is based on the electrical nature of corrosion and is usually applied to water pipelines.

Figure 4. Coatings for protection against corrosion pipelines.

14.4 Corrosion Inhibitors

Corrosion can be controlled by modifying the water environment and by neutralising or removing corrosive agents (e.g., DO). Corrosion inhibitors slow the rate of corrosion reactions when added in relatively small amounts to the water. They are divided into three groups:

- anodic inhibitors, which retard the anodic corrosion reactions by forming passive films;
- cathodic inhibitors, which repress the corrosion reaction such as reducing DO;
- adsorption inhibitors, such as amines, oils, and waxes, which are adsorbed on the steel surface to form a thin protective film that prevents metal dissolution.

These conventional inhibitors are applied in many sectors of the water and energy industries—cooling water systems,[3,8-9] desalination plants, [10-12] coal water slurries,[3] acid pickling of metals,[3] on reinforcing steel in concrete, and for control of galvanic corrosion in heat exchangers exposed to reverse osmosis water (table 1).[12]

Table 1. Applied corrosion inhibitors for water systems[A]

System	Corrosion inhibitors
Engine coolants	molybdate
	molybdate with nitrite; molybdate, arsenite, or arsenate and benzotriazole along with borate/phosphate/amine
	Nitrite, nitrate, phosphate, borate, silicate, benzoate, aminophosphonate, phosphinopolycarboxylate, polyacrylate, hydroxybenzoate, phthalate, adipate, benzotriazole, tolytriazole, mercaptobenzothiazole, and triethanolamine are combined with molybdate. In glycol, 0.1 to 0.6 wt% of molybdate is used.
Closed recirculating cooling water	200 ppm sodium molybdate with 100 ppm of sodium nitrite
	50 ppm molybdate, 50 ppm phosphate, 2 ppm Zn^{2+}
	40 ppm sodium molybdate + 40 ppm sodium silicate
	2-phosphonobutane-1,2,4-tricarboxilic acid and polyvinylpyrrolidone
Cooling water of steam plant boiler waters	molybdate with an aluminium salt and thiourea
	mild steel corrosion inhibition in boilers by a mixture of sodium molybdate, sodium citrate, manganese sulphate, polymaleic acid, and morpholine
	protection of mild steel in hard water boilers by sodium molybdate and sodium nitrite

[A]Source: V. S. Sastri, Green Corrosion Inhibitors: Theory and Practice, Wiley and Sons (2011).

In the last decade, a new family of inhibitors has emerged called green corrosion inhibitors, which are relevant in this crucial time of energy problems and economic havoc since they will extend the life of the water infrastructure and save large expenses in materials, equipment, and structures. They belong to the advanced field of green chemistry, also known as sustainable chemistry, which involves the design of chemical products and processes that reduce or eliminate the use or generation of hazardous substances.

In one example, R. Garcia[13] evaluated the inhibitive action of an ethanol extract derived from a desert plant on the corrosion of CS in hydrochloric acid (HCl) and found it to be effective. HCl is employed at times to remove carbonate scales from steel surfaces.[13-14]

14.5 Conclusions

Corrosion is a damaging process that affects the water infrastructure including pipelines, pumps, valves, and auxiliary equipment. Economic considerations are of the utmost importance when evaluating anticorrosion methods involving inhibitors. The use of an inhibitor under operating conditions is determined largely by both its chemical stability and its corrosion prevention efficiency. Conventional corrosion inhibitors, especially green corrosion inhibitors, will contribute to maintain effective water systems and their related natural and manmade environments.

14.6 References

1. B. Valdez, M. Schorr, et al., 'Effect of Climate Change on the Durability of Engineering Materials in Hydraulic Infrastructure: An Overview,' Corr. Eng. Sci. and Technol. 45, 1 (2010): pp. 34–41.

2. B. Valdez, M Schorr, eds., 'Special Issue: Relationship of Corrosion with Climate Change,' Corr. Eng. Sci. Technol. 45 (2010).

3. V. S. Sastri, Green Corrosion Inhibitors: Theory and Practice (Hoboken, NJ: John Wiley and Sons, 2011), pp. 212, 216, 223.

4. W. Sung, 'Corrosion in Potable Water Distribution and Building Systems,' S. D. Cramer, B. S. Covino Jr., eds., Corrosion: Environments and Industries, ASM Handbook, Vol. 13C (Materials Park, OH: ASM International, 2006), pp. 8–11.

5. Corrosion of Metals, Wear and Erosion, Annual Book of ASTM Standards, Vol. 03.02 (West Conshohocken, PA: ASTM, 2012).

6. NACE TM0169-2000, 'Laboratory Corrosion Testing of Metals' (Houston, TX: NACE International, 2012).

7. P. Dupont, J. P. Peri, 'World Class Water Pumps,' Sulzer Technical Review 3 (2011): pp. 12–15.

8. B. P. Boffardi, 'Corrosion Inhibitors in the Water Treatment Industry,' S. D. Cramer, B. S. Covino Jr., eds., Corrosion: Fundamentals, Testing and Protection, ASM Handbook, Volume 13A (Materials Park, OH: ASM International, 2003).

9. A. Abulkibash et al., 'Corrosion Inhibition of Steel in Cooling Water System by 2-Phosphonobutane-1,2,4-Tricarboxilic Acid and Polivinylpyrrolidone,' The Arabian J. for Sci. and Eng. 33 (1A), 1 (2008): pp. 29–40.

10. M. Schorr, B. Valdez, J. Ocampo, A. Eliezer, 'Corrosion Control in the Desalination Industry,' M. Schorr, ed., Desalination, Trends and Technologies (New York, NY: Intech, 2011).

11. M. Schorr, B. Valdez, J. Ocampo, A. Eliezer, 'Materials and Corrosion Control in Desalination Plants,' MP 51,5 (2012): pp. 56–60.

12. I. Carrillo, B. Valdez, M. Schorr, R. Zlatev, 'Inorganic Inhibitors Mixture for Control of Galvanic Corrosion of Metals Cleaning Process Industry,' CORROSION 2012 (Houston, TX: NACE, 2012).

13. R. Garcia, B. Valdez, R. Kharshan, A. Furman, M. Schorr, 'Interesting Behaviour of Pachycormus Discolor Leaves Ethanol Extract as a Corrosion Inhibitor of Steel in 1 M HCl: A Preliminary Study,' Intl. J. of Corrosion (2012).

14. C. Chandler, M. Kharshan, A. Furman, 'Sugar Beets against Corrosion,' Corrosion Reviews 20, 4-5 (2002): pp. 379–390.

CHAPTER 15

Copper Corrosion by Atmospheric Pollutants in the Electronics Industry

Benjamin Valdez Salas,[1] Michael Schorr Wiener,[1] Roumen Zlatev Koytchev,[1] Gustavo López Badilla,[2] Rogelio Ramos Irigoyen,[1] Monica Carrillo Beltrán,[1] Nicola Radnev Nedev,[1] Mario Curiel Alvarez,[1] Navor Rosas Gonzalez,[2] and Jose María Bastidas Rull[3]

[1]Engineering Institute, Autonomous University of Baja California, Boulevard Benito Juarez y Calle a la Normal S/N, Colonia Insurgentes Este, 21280 Mexicali, BCN, Mexico
[2]Polytechnic University of Baja California, Calle de la Claridad S/N, Colonia Plutarco Elias Calles, 21376 Mexicali, BCN, Mexico
[3]National Centre of Metallurgical Research, Avenue Gregorio del Amo 8, 28040 Madrid, Spain

Hydrogen sulphide (H_2S) is considered one of the most corrosive atmospheric pollutants. It is a weak, diprotic, reducing acid, readily soluble in water and dispersed into the air by winds when emitted from natural, industrial, and anthropogenic sources. It is a pollutant with a high level of toxicity, impairing human health and the environment quality. It attacks copper, forming thin films of metallic sulphides or dendrite whiskers, which are cathodic to the metal substrate, enhancing corrosion. H_2S is actively involved in microbially influenced corrosion (MIC), which develops in water, involving sulphur-based bacteria in oxidising and reducing chemical reactions. H_2S is found in concentrated geothermal brines, in the atmosphere of geothermal fields, and in municipal sewage systems. Other active atmospheric pollutants include SO_x, NO_x, and CO. This investigation reports on the effects of H_2S on copper in microelectronic components of equipment and devices, with the formation of non-conductive films that lead to electrical failures.

15.1 Introduction

The electronics industry is spread out worldwide; it is an important sector in the Mexican economy, representing 80 per cent of industrial companies in the northwest of the country. Their assembly plants are located in three cities: Mexicali, an arid zone; Tijuana, an urban-industrial area; and Ensenada, a marine region on the Pacific Ocean coast, all belonging to the state of Baja California, near the Mexico-USA border. The electronics industry appeared in Mexico during the '60s with the manufacture of electronic products such as radios, audio record and play devices, and televisions. This industry designs and manufactures microelectronic components called microcontroller devices (MCD), integrated with microelectromechanical systems (MEMS). A study was conducted in the indoor areas of three electronics plants in these cities. Copper and its alloys are widely applied in the electric energy, electronics, and semiconductor industries because of their high electrical and thermal conductivity, ductility, and malleability.

Copper is considered a noble metal; it resists attack by oxygen, although some air pollutants, such as H_2S, change its surface properties, even at ambient temperature, forming a thin layer having completely different properties compared with the pure metal surface. This layer lowers catastrophically the adhesion of the soldering alloy or conductive resins and paste, provoking failures of the printed circuit board (PCB) of the microelectronic devices. Compounds such as geerite (Cu_8S_5) are formed on Cu in the presence of H_2S and patinas as Cu_2O; posnjakite ($Cu_4SO(OH)_6H_2O$), brochantite ($Cu_4SO_4(OH)_6H_2O$), and antlerite ($Cu_3SO_4(OH)_4$) are formed in the presence of humidity.[1] The formation of tarnish films on a copper surface exposed to environments containing atmospheric pollutants and high humidity involves the movement of metallic ions over the surface, away from the metal, generating a creep process that increases the contact resistance, leading to electric failures of the electronic devices. Copper sulphidation is a fast process occurring on the metal-gas phase interface, impairing the Cu corrosion resistance.[2,3] This paper presents the corrosion process of Cu exposed to H_2S polluted environments, under varied conditions of temperature and humidity, in the indoor areas of electronics manufacturing plants located in Mexicali, Baja California, Mexico.[4]

15.2 H_2S, a Corrosive, Toxic Pollutant

It is appropriate to report in the context of the present paper about the corrosivity and toxicity of H_2S since this also affects the quality of the environment, human health, and the durability of the engineering materials, which are central issues of modern society. H_2S acts as a pollutant in the indoor areas of manufacturing plants of the electronics industry; it promotes the formation of thin copper sulphide films on PCB surfaces. Recent investigations[5,6] proved that the main H_2S source is a geothermal field generating underground sources of steam and H_2S located about 40 km south from Mexicali City. To avoid this air pollution and consequent corrosion is impossible without the application of high-cost air cleaners.[7,8]

H_2S gas emitted into the atmosphere from additional heavy sources such as municipal sewage causes respiratory diseases and inflammation of the eyes; it has an offensive odour of rotten eggs; therefore, it is easy to detect, even at low concentration of 10 to 30 ppb (parts per billion) in the atmosphere around the geothermal fields.

The corrosion activity of H_2S is evident in

$$Fe + H_2S \rightarrow FeS + H_2, \qquad (1)$$

$$Cu + H_2S \rightarrow CuS + H_2. \qquad (2)$$

15.3 Experimental

Air pollutants measurements. Data on air pollutants were gathered every five minutes and organised in files for monthly periods. The specialised instruments controlled by the United States Environmental Protection Agency (US-EPA) monitoring air pollution were a chemiluminescence NO_x analyser, model 42 of Thermo Environmental Instruments Inc., a gas filter CO analyser model 3000E of Advanced Pollution Instruments Inc. (API), a SO_2 photometric analyser from Thermo Electron Corporation, and an O_3 analyser model 400 of API. This electronic instrument has filters to separate dust particles from gases.

The practices recommended in ISO standards for atmospheric corrosion were taken into account. The Cerro Prieto geothermal wells, in the vicinity of Mexicali, emit H_2S into the atmosphere surrounding the fields and the power plants. Other H_2S emissions came from the plant chimney stacks, vapour ducts, noise silencers, and cooling towers, totalling 22,740 t per year.[9,10] Since the production capacity of Cerro Prieto has not changed over the last five years or so, it may be estimated that the same H_2S concentration in the atmosphere, around the geothermal wells, ranges from 10 to 30 ppb. Typical ranges of natural and anthropogenic H_2S under outdoor and indoor conditions are 0.724 ppb and 0.1 to 0.7 ppb, respectively.[11, 12]

Corrosion rate measurement. Rectangular metallic specimens of Cu with an exposition surface of 6.45 cm^2 were prepared. The specimens were polished to 400 SiC paper, washed, degreased with acetone, dried with hot air, and weighed before being installed in a metallic chamber of exposure under indoor conditions for one-, three-, six-, twelve- and twenty-four-month periods. The corrosion rates were determined by applying the gravimetric method according to ASTM G31 standard method. To simulate controlled indoor conditions, the chamber was fabricated with pre-coated aluminium with a total volume of 0.1 m^3 and conditioned with two air inlet blinds coupled to metallic filters to permit the penetration of gases with the flow of air, to prevent the penetration of dust, and to avoid mistakes in the weight loss calculations.[13, 14] The chamber was provided with metallic internal supports to hold the specimens; it was installed on the roof of an electronics manufacturing plant at 10 m above ground level.[15, 16]

After each period of exposure, the metallic samples were removed and weighed to obtain the mass gain. The corrosion products morphology was observed with a stereoscope before being removed, cleaned, and reweighed to obtain the mass loss on an analytical balance to the nearest 0.00001 g of accuracy. Corrosion of Cu surfaces under constant concentration of H_2S and controlled relative humidity (RH) conditions occurred using a closed system consisting of an acrylic sealed chamber with an inlet valve to provide a 0.1 ppm H_2S concentration and a ventilator coupled to a humidity generator to provide 80 per cent RH.[17, 18]

Surface examination. The copper specimens exposed to a controlled H_2S environment during short times were analysed to determine their surface characteristics by a scanning electron microscope (SEM), applying a JEOL (JEOL Ltd., Peabody, MA) JSM-6360, coupled with an Energy Dispersive X-ray (EDX) analysers (AMETEK Inc., Mahwah, NJ), used for chemical composition analysis.

15.4 Results

Corrosion of copper. The corrosion rate (CR) of Cu specimens exposed in the test chamber during twenty-four months (figure 1) demonstrates that the extent of corrosion augments with the exposure time reaching values of 270 mg·m^{-2} after the two years of exposure at RH values ranged from 15 per cent to 75 per cent and temperatures from 4°C to 45°C, depending on the year season; SO_2 and NO_x were the species with major concentration levels.

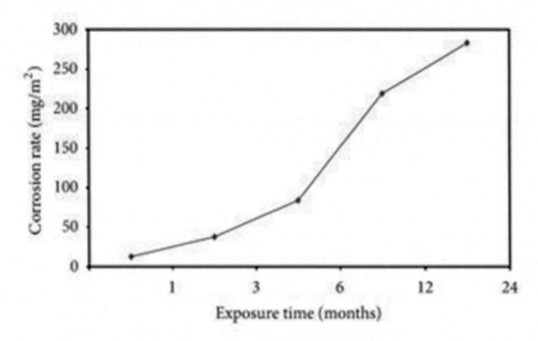

Figure 1. Corrosion rate of copper exposed in the corrosion chamber during twenty-four months.

Table 1. Copper sulphides in films formed by reaction with H_2S at indoor areas of electronics plants in Mexicali City

Mineral	Chemical composition
Chalcocite	Cu_2S
Djurlcite	$Cu_{1.95}S$
Digenite	$Cu_{1.82}S$
Geerite	$Cu_{1.62}S$

Corrosion of Cu in the presence of air, H_2S, and humidity produces oxides and/or sulphides as wet films leading to electrical failures in electronic equipment.[10] On the other hand, corrosion of Cu occurs when the RH overpasses the 80 per cent and the SO_2 concentration is larger than 0.1 ppm. The corrosion behaviour also depends on the state of the metal surface: smooth, corrugated, polished, and non-uniform. Sometimes, a Cu_2O film that slows the rate of corrosion gradually dissolves in the presence of an acidic electrolyte constituted by SO_2, and then the corrosion products formed are different because of the surface conditions. At levels of SO_2 greater than the air quality standards, in combination with NO_2 and O_3 at different concentrations in indoor plants, cuprite (Cu_2O) and copper sulphides form, as depicted in figures 2 and 3.

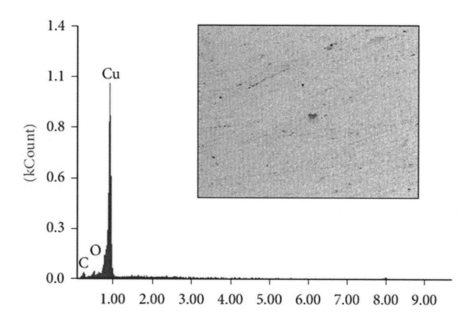

Figure 2. SEM picture for Cu surface after two days in 0.3 ppm H$_2$S/air, magnification 400x.

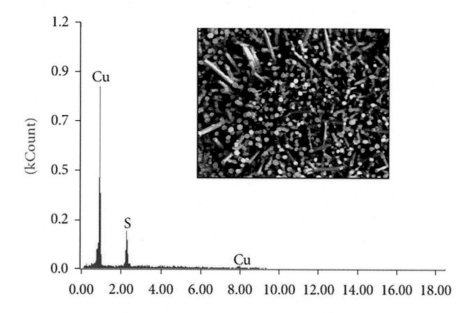

Figure 3. SEM picture for Cu surface after two days in 0.3 ppm H$_2$S/ air, magnification 1200x.

Some spectacular structures of hexagonal crystals of Cu sulphide are shown in figure 3. The SEM results of the exposed samples showed the formation of spots of CuS during the first two days, which grew continuously until all spots united together as large films covering the entire sample surface. The concentration of H$_2$S in the electronics plant atmosphere was 0.9 ppm; at the Cerro Prieto geothermal field atmosphere, it reaches 1.5 ppm and higher concentrations because of the continuous emission of gases accompanying the produced steam.[11]

The EDX analysis performed in different points of the corroded copper surfaces reveals the formation of several sulphides of stoichiometric composition with a general formula CuS, where the values of vary from

1.6 to 2. The estimated composition of the copper sulphide corrosion products and the corresponding mineral are displayed in table 1.

Film formation mechanisms. Under conditions of humidity and in contact with air contaminated with H_2S, Cu generates two types of corrosion products: oxides and sulphides, according to the following electrochemical reactions.

Oxidation

$$2Cu \rightarrow 2Cu^{2+} + 4e^- \tag{3}$$

Oxidation, anodic reaction

$$O_2 + 2H_2O + 4e^- \rightarrow 4OH^- \tag{4}$$

Reduction, cathodic reaction

$$2Cu + O_2 + 2H_2O \rightarrow 2Cu(OH)_2 \tag{5}$$

Total corrosion reaction

$$2Cu(OH) \rightarrow Cu_2O + H_2O \tag{6}$$

Hydroxide converts to oxide

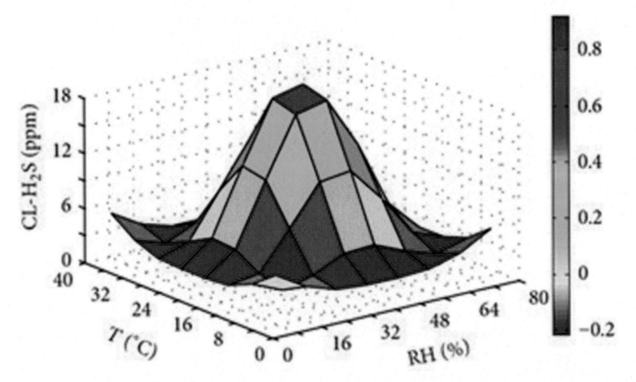

Figure 4. Correlation of copper corrosion with H_2S, relative humidity %, and temperature in the electronics industry (summer, 2010).

Sulphidation

$$Cu \rightarrow Cu^{2+} + 2e^- \qquad (7)$$

Oxidation, anodic reaction

$$H_2S \rightarrow H^+ + SH^- \qquad (8)$$

Reduction, cathodic reaction

$$2SH^- \rightarrow 2S^- + H_2 \qquad (9)$$

Reduction, cathodic reduction

$$Cu + H_2S \rightarrow CuS + H_2 \qquad (10)$$

Total sulphidation reaction

Influence of atmospheric pollutants. Mexicali City has highly contaminated air because of the presence of fine dust coming from the desert around, but the gaseous pollutants such as SO_x, NO_x, CO, and O_3 are generated from the diverse industrial activities. Air pollutants penetrate to indoor locations of electronics plants and corrode Cu-made devices. The relative humidity (RH) and temperature reach up to 50 per cent and 30°C during the major part of the year. Table 2 presents the relation between the concentrations of these pollutants and the Mexicali climatic factors.

Table 2. Relation of concentration of air pollutants and climate factors in Mexicali

| Seasons | Climatic factors | | | | | | | | | | | |
| | Sulphur dioxide (SO_2) | | | Carbon monoxide (CO) | | | Nitrogen oxides (NO) | | | Ozone (O_3) | | |
	RH[a]	T[b]	C[c]	RH[a]	T[b]	C[c]	RH[a]	T[b]	C[c]	RH[a]	T[b]	C[c]
Spring												
Max	49.3	34.8	0.16	39.1	29.3	69	34.3	28.6	0.48	48.3	27.3	0.21
Min	23.3	20.4	0.08	28.6	22.1	4	29.8	19.9	0.01	28.5	15.2	0.06
Summer												
Max	83.8	45.9	0.11	73.9	45.9	57	70.8	39.9	0.54	73.2	44.5	0.37
Min	46.1	23.6	0.03	47.5	29.2	16	44.7	22.4	0.18	44.3	26.5	0.01
Winter												
Max	77.8	27.4	0.50	72.8	24.7	84	69.1	24.6	0.75	87.9	27.7	0.51
Min	16.6	17.8	0.17	39.3	23.9	6	63.2	13.8	0.17	47.2	28.8	0.05

[a]RH: relative humidity, %; [b]T: temperature, °C; and [c]C: air pollution concentration (C), ppm.

The two most aggressive pollutants are sulphur-containing H_2S and SO_2, both acidic, but one reducing (H_2S) and the other oxidant (SO_2) agents. Their behaviour and corrosivity during the seasons of the year are presented in table 3.

Table 3. Correlation of corrosion rate, the year season, and air pollutants in indoor conditions of industrial plants

Seasons	Climatic factors							
	Hydrogen sulphide (H_2S)				Sulphur dioxide (SO_2)			
	RH[a]	T[b]	C[c]	CR[d]	RH[a]	T[b]	C[c]	CR[d]
Spring								
Max	88.8	33.4	0.15	255	85.6	23.2	0.34	176
Min	34.5	17.6	0.09	130	46.7	15.1	0.23	112
Summer								
Max	89.9	42.1	0.14	265	88.2	39.9	0.45	245
Min	38.5	24.3	0.11	181	42.3	28.2	0.18	114
Winter								
Max	87.5	25.6	0.42	382	88.8	22.3	0.67	338
Min	43.2	17.8	0.26	245	38.9	12.3	0.25	136

[a]RH: relative humidity, %; [b]T: temperature, °C; [c]C: air pollution concentration (C), ppm; and [d]CR: corrosion rate, mg/m². year

The corrosion data collected, arranged in tables 2 and 3, correlate values of RH, temperature, and CR provoked by the different atmospheric pollutants. These data were evaluated and displayed using the MATLAB software, a mathematical computing software (MathWorks Inc., USA), to determine the relationship between the environmental factors and the corrosion rate of metals used in the electronics industry. Figure 4 is a particular 3D graph depicting the correlation of the CR of Cu with the RH and the temperature, indicating in circles the maximum and the minimum CR.

The maximum CR appears at 88 per cent RH and 16°C, and the minimum CR is recorded at 18 per cent RH and 2.0°C, expressing the critical influence of humidity and temperature levels. These levels are controlled inside the electronics plant, but sometimes corrosion occurs.

An additional MATLAB graph depicts the influence of frequent industrial pollutants: NO_x, SO_x, O_3, and CO_x on the corrosivity indexes of Cu published by ISO, the International Organization for Standardization[7] (figure 5).

Correlation of corrosivity levels with atmospheric pollution

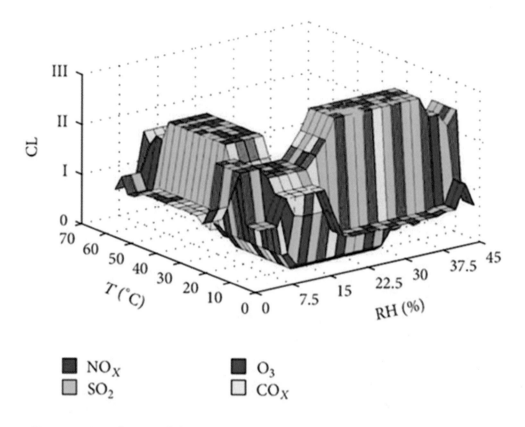

Figure 5. Correlation of climatic factors and pollutants with corrosivity indexes.

Other MATLAB graphs correlate the CR of Cu with the climatic factors and pollutants in summer 2010 in Mexicali (figure 6) and those in winter 2010 too (figure 7).

Correlation of CR with climatic factors in summer

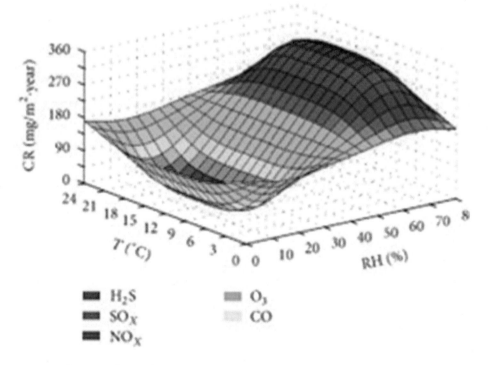

Figure 6. Correlation of CR of copper with climatic factors in summer in Mexicali (2010).

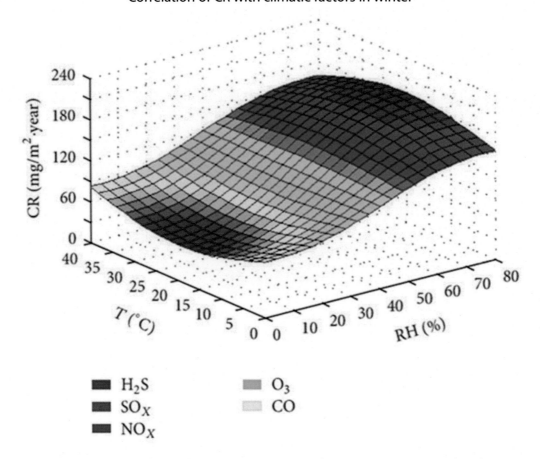

Figure 7. Correlation of CR of copper with climatic factors in winter in Mexicali (2010).

15.5 Discussion

The climatic variables and the atmospheric pollutants are the principal factors that enhance the corrosion in indoor conditions of the metals utilised in the electronics industry of the state of Baja California, Mexico. The evaluation of these parameters and their effect on the metals' surfaces demonstrates the relationship of atmospheric corrosion with the damage caused to the electrical connections, resistors, diodes, connectors, and wires of the electrical-electronic equipment. This corrosion damage generates low yielding by electrical failures in the industrial devices and equipment. The maximum and minimum RH, temperature, and CR and the relationship with the air pollutants were analysed during different seasons of the year. These data were expressed graphically applying the MATLAB software. A study on the correlation of climate factors with the function of electronic test equipment installed inside a clean room of an electronics plant in the city of Tijuana was conducted; it was done at various levels of RH and temperature, relating them to the electric current circulating in the test equipment to indicate the correct or incorrect state of the microcomponents. They include electronic components such as transistors, capacitors, coils, resistors, and diodes that are assembled in the semiconductor wafer based on a silicon microelectronic board. The bad air quality in Tijuana should be attributed to the last cold winters that have caused more burning of fossil fuels to supply electricity for urban-heating systems, increasing the emissions of corrosive pollutants, worsening the smog, which comprises fire smoke, car exhaust gases, and dust, all trapped in the air mist. The health impact is vast too, raising the number of people suffering from respiratory illness that triggers heart and asthma attacks. The city of Ensenada is located on the coast of the Pacific Ocean, in the northwest of Mexico in a marine region. It has a tropical marine climate with cold winter mornings around 5°C and 35°C in the summer; RH is around 20 per cent to 80 per cent, varying during the seasons of the year. The climate factors analysed were humidity, temperature, winds, and rains to determine the time of wetness (TOW), a critical factor in the determination of CR and its extent.

Santa Ana winds (SAW) constitute a climatic phenomenon that alters the atmospheric conditions; they originate in the Santa Ana canyon, in the Mojave desert, which cause fast changes in the climate conditions in southwest California and northwest Baja California. SAW form when the desert becomes cooler, usually during the autumn and spring seasons; rising temperatures, humidity, and meteorological conditions influence the indoor environment of the electronics industry. The principal air corrodent encountered in Ensenada is the NaCl aerosols from the sea, in addition to CO, NO_x, and SO_2 from traffic vehicles, power stations, and industrial and landfill emissions, which increase atmospheric corrosivity.[13, 14]

15.6 Conclusions and Recommendations

A study was conducted during a span of two years on the corrosion of metallic materials used in the manufacture of electronic devices and equipment. The gaseous air pollutants—for example, H_2S, SO_x, NO_x, and CO, generated by the geothermal fields, the electricity industry, and the motor vehicles burning fossil fuels—lead to the appearance of corrosion on the metals' surfaces. Copper suffers in particular because of the attack by sulphur-containing pollutants H_2S and SO_x, forming copper sulphides and oxides that impair their electrical conductivity properties. The installation and effective maintenance systems, to clean and control the contaminated air that infiltrates into the electronics production rooms such as filters, would prevent and/or minimise this atmospheric corrosion. Encapsulation and hermetic sealing of microcomponents prevent a reaction with the pollutants. Removal of moisture by efficient air conditioning and continuous maintenance of an optimum environmental condition of the indoor areas of electronics plants avoids and/or mitigates corrosion.

15.7 References

1. B. Valdez, M. Schorr, M. Quintero, et al., 'Corrosion and scaling at Cerro Prieto geothermal field,' Anti-Corrosion Methods and Materials, vol. 56, no. 1, pp. 28–34, 2009.
2. L. Veleva, B. Valdez, G. Lopez, L. Vargas, and J. Flores, 'Atmospheric corrosion of electro-electronics metals in urban desert simulated indoor environment,' Corrosion Engineering Science and Technology, vol. 43, no. 2, pp. 149–155, 2008.
3. J. F. Flores and S. B. Valdez, 'Cabina de simulacion´ de corrosion para la industria electronica´ en interior,' Ingenieros, vol. 6, no. 21, 2003 (Spanish).
4. ASTM, 'Standard practice for conducting atmospheric corrosion test on metals,' ASTM G50-76, ASTM, West Conshohocken, Pa, USA, 2003.
5. G. Lopez, B. Valdez, and M. Schorr, 'Spectroscopy analysis of corrosion in the electronics industry influenced by Santa Ana Winds in marine environments of Mexico,' in Indoor and Outdoor Air Pollution, INTECH, 2011.
6. 'Corrosion of metals and alloys. Classification of low corrosivity of indoor atmospheres: determination and estimation attack in indoor atmospheres,' ISO 11844-1, ISO, Geneva, Switzerland, 2005.
7. 'Corrosion of metals and alloys. Classification of low corrosivity of indoor atmospheres: determination and estimation of indoor corrosivity,' ISO 11844-2, ISO, Geneva, Switzerland, 2006.
8. A. Moncmanova, Environmental Deterioration of Materials, WIT Press, 2007.

9. L. B. Gustavo, Caracterizacion´ de la corrosion´ en materiales metalicos´ de la industria electronica´ en Mexicali, B.C. [Tesis de doctorado], UABC, Instituto de Ingenier´ıa, Mexicali, Mexico,´ 2008.

10. G. Lopez, H. Tiznado, G. Soto, W. de la Cruz, B. Valdez, and R. M. Schorr Zlatev, 'Corrosion´ de dispositivos electronicos´ por contaminacion´ atmosferica´ en interiores de plantas de ambientes aridos´ y marinos,' Revista Nova Scientia, vol. 3, no. 1, 2010.

11. G. Lopez, H. Tiznado, G. S. Herrera, et al., 'Use of AES in corrosion of copper connectors of electronic devices and equipment's in arid and marine environments,' Anti-Corrosion Methods and Materials, vol. 58, no. 6, pp. 331–336, 2011.

12. M. Reid, J. Punch, C. Ryan, et al., 'Microstructural development of copper sulfide on copper exposed to humid H_2S,' Journal of the Electrochemical Society, vol. 154, no. 4, pp. C209–C214, 2007.

13. B. Valdez, M. Schorr, R. Zlatev et al., 'Corrosion control in industry,' in Environment and Industrial Corrosion, Practical and Theoretical Aspects, INTECH, 2012.

14. S. B. Valdez, W. M. Schorr, B. G. Lopez, et al., 'H_2S pollution and its effect on corrosion of electronic components,' in Air Quality-New Perspective, INTECH, 2012.

15. B. G. Lopez, S. B. Valdez, W. M. Schorr, and G. C. Navarro, 'Microscopy and spectroscopy of MEMS used in the electronic industry of Baja California region Mexico,' in Air Quality-New Perspective, INTECH, 2012.

16. B. G. Lopez, S. B. Valdez, K. R. Zlatev, P. J. Flores, B. M. Carrillo, and W. M. Schorr, 'Corrosion of metals at indoor conditions in the electronics manufacturing industry,' Anti-Corrosion Methods and Materials, vol. 54, no. 6, pp. 354–359, 2007.

17. J. Smith, Z. Qin, F. King, L. Werme, and D. W. Shoesmith, 'Sulfide film formation on copper under electrochemical and natural corrosion conditions,' Corrosion, vol. 63, no. 2, pp. 135–144, 2007.

18. K. Demirkan, G. E. Derkits Jr., D. A. Fleming, et al., 'Corrosion of Cu under highly corrosive environments,' Journal of the Electrochemical Society, vol. 157, no. 1, pp. C30–C35, 2010.

16

Optimization and Characterisation of Commercial Water-Based Volatile Corrosion Inhibitor

N. Cheng,[1] B. Valdez-Salas,[2] P. Moe,[1] and J. S. Salvador-Carlos[2]

[1]Magna International Pte Ltd., 10 H Enterprise Road, 629834, Singapore
[2]Materiales Avanzados y Corrosion, Instituto de Ingeniería de UABC, Mexicali Calle Normal s/n, Parcela 44, 21100 Mexicali, Mexico

16.1 Abstract

Volatile corrosion inhibitor (VCI) provides protection for metal surfaces. VCI coating and molecules attach themselves to metal surfaces to form both a physical film when contacted and an invisible thin film for indirect contact through vapour (only a few molecules thick), thus inhibiting metals atmospheric corrosion. Optimisation and characterisation of commercial water-based volatile corrosion inhibitor VCI (Vapour-Phase-Protection, VAPPRO 837C diluted commercial solution from CORPPRO) was prepared to determine their characteristics and effectiveness against corrosion of carbon steel. The main scope of this work is to characterise the rheological and corrosion inhibition properties of the VAPPRO 837C with varying formulations and processing parameters (coating and drying times). Different tests were performed to determine the corrosion behaviour of inhibitor. The application of VCI on the metal surface was done by dip-coating process. An electrochemical workstation from HCH Instruments has been used to evaluate the corrosion inhibition efficiency of the VCI and to determine the corrosion rate of the uncoated and coated samples. In addition, viscosity tests were carried out to determine the rheological properties of the formulation, as well as freeze-thaw resistance of waterborne coatings and pH tests were done. An FTIR spectrometer has been used to determine the functional groups present at a specific concentration. Results reveal that the most effective VCI film was obtained from a 0.25 CORPPRO (concentrated VCI) (vol%) formulation using a coating time of 10 min and a drying time of 24 h. Therefore, 0.25 CORPPRO would be the optimum concentration to be used because it is able to achieve the highest corrosion inhibition efficiency with the optimum coating drying time.

Keywords: corrosion inhibitor, vapour phase protection, dip-coating

16.2 Introduction

Metallic corrosion is a worldwide problem caused by the interaction of the metal surface with the surrounding environment. It has always been problematic to metal constructs and tools as degradation or rusting of metals because of the tendency to return to their natural state (oxides or other corrosion leading to a weakening of mechanical properties and failure).[1] Replacing them outright is expensive and, with resources on earth becoming more limited, problematic in the long run.[2] Thus, corrosion protection of metallic infrastructures is preferable by their cost over part replacement and prolongs the working life of the product in the field, with many methods to do so currently available such as corrosion inhibitors, coatings, and cathodic protection.[3]

Volatile corrosion inhibitors are widely used as corrosion control and prevention method. They are a part of a class of corrosion-inhibiting compounds with a finite vapour pressure, where the inhibitors are transported to the target metal through space to condense on the metal surface, forming a protective film, lowering the corrosion rate of the metal itself.[4] Factors that determine the efficiency of VCIs are the concentration of VCI compounds, the period of exposure, and vapour pressure.[5] The main advantage of these inhibitors is their user-friendly application of VCI onto the metal. Other benefits include to reach crevices, blind holes, and other 'difficult to reach' areas. Limitations include being temporary films as they can be removed easily, are used at fairly low concentrations and a higher concentration is needed for the self-healing effect, which could increase the cost of using it and corrosion protection may not be the only requirement, with others such as colour limitation and film hardness needed. It is due to these factors that they are largely ignored in industrial maintenance coatings.[6] However, alongside corrosion-resistant materials and corrosion protection coatings, corrosion inhibitors are still researched and developed to further lower corrosion rates and to reduce the costs of corrosion.[7]

An example of these inhibitors is the ammonia. This compound has been used as a VCI to protect immersed and exposed parts of steam boiler circuits at the beginning of the twentieth century.[8] In the 1940s, less odorous, safer, and more effective substances were used for protection. With more than one thousand types of VCI compounds known today, only a few are used as these are acceptably efficient, cost effective, and environmentally friendly.[9] One type of inhibitor used in VCIs is salts of dicyclohexylamine such as dicyclohexylammonium nitrite (DCHN) and cyclohexylamine carbonate (CHC).[10] The VCI developed by Magna International uses amine carboxylate as the inhibitor. It is a salt synthesised by neutralising carboxylic acid with a blend of amines, with the number of carbon atoms ranging from 1 to 26, with corrosion inhibition increasing with the number of carbons. Other than corrosion inhibition, it is also used for boundary lubrication, emulsification, and detergency.[11]

Recently, there has been a shift in using water-based VCIs, which are less hazardous to human health and are environmentally friendly, as opposed to oil-based VCIs. By substituting the organic solvent with water as the transport medium, it would lower the levels of volatile organic compounds (VOCs) emitted by VCIs.[12] This study aims to deduce the concentration of the VCI developed by Magna International, VAPPRO 837C, that can offer the best corrosion protection and under what conditions can this be achieved. The effects of the concentration of inhibitors, coating time, drying time, and temperature on its performance are studied and observed.

16.3 Experimental

16.3.1 Preparation of Metal Samples

Carbon steel UNS G10100 disks with 0.7853 cm^2 surface area were prepared and de-rusted by hand using 600 grit abrasive paper. Last, the samples were raised with deionised water and isopropyl alcohol to remove debris from the de-rusting procedure.[13]

16.3.1 Determination of Corrosion Rate and Corrosion Inhibitor Efficiency

Electrochemical corrosion tests were performed using an HCH Instruments electrochemical workstation and a three-electrode cell array. The electrochemical method was DC polarisation using the coated and uncoated samples as working electrode, and a silver/silver chloride (Ag/AgCl) as reference electrode, while a platinum mesh was used as counter electrode. A schematic diagram of the electrochemical corrosion cell is given in figure 1.

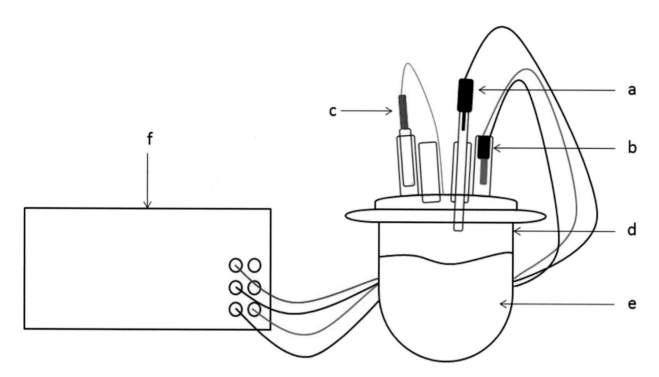

Figure 1. Schematic diagram of three-electrodes cell: (a) reference electrode (Ag/AgCl), (b) working electrode (sample), (c) counter-electrode (platinum Pt), (d) electrolyte container, (e) electrolyte, and (f) potentiostat.

The parameters established for the electrochemical test were 10 mV s^{-1} scan rate and the polarisation range from –100 to –600 mV versus corrosion potential. Final results were adjusted by Tafel analysis to calculate corrosion rates and the percentage of corrosion inhibition efficiency. The corrosive media was Singapore tap water with the typical values of microbiological and physicochemical parameters shown in table 1.[14]

The corrosion rate was calculated using the following formula:

$$Corrosion\ rate = k[\frac{(a.i_{corr})}{(n.D)}] \qquad (1)$$

where D is density of metal specimen, a is atomic weight, i_{corr} is corrosion current density, n is the number of electrons lost and k is the constant, which depends on unit of corrosion rate. For this work, the k value applied was 0.00327 mm/(A cm year).

Table 1. Microbiological and physical chemical parameters typical values

Parameter	Unit	Typical value
Microbiological		
Escherichia coli (E. coli)	cfu/100 mL	< 1
Physical		
Colour	Hazen	< 5
Conductivity	µS/cm	< 250
Chlorine	mg/L	< 2.0
pH	Units	7.0 – 8.5
Total dissolved solids (TDS)	mg/L	< 150.0
Turbidity	NTU	< 5
Chemical		
Ammonia (as N)	mg/L	< 1.0
Calcium	mg/L	4.0 – 20.0
Chloride	mg/L	< 20.0
Copper	mg/L	< 0.05
Fluoride	mg/L	< 0.50
Iron	mg/L	< 0.04
Manganese	mg/L	< 0.05
Nitrite (as N)	mg/L	< 11.0
Sodium	mg/L	< 20.0
Sulphate	mg/L	< 5.0
Silica (as SiO_2)	mg/L	< 3.0
Total organic carbon (TOC)	mg/L	< 0.50
Total hardness (as $CaCO_3$)	mg/L	< 50.0
Zinc	mg/L	< 0.10

Based on the corrosion rates of the metal sample before and after coating, the corrosion inhibition efficiency was calculated based on the formula below:[15]

$$\text{Corrosion inhibition efficiency}(\%) = \frac{(CR_{uninhibited} - CR_{inhibited})}{CR_{uninhibited}} \times 100 \tag{2}$$

where $CR_{uninhibited}$ is the corrosion rate of uninhibited sample and $CR_{inhibited}$ is the corrosion rate of inhibited sample.

16.3.3 Formulation of VAPPRO 837

A concentrated corrosion inhibitor solution based on amine carboxylates was used to prepare formulations of commercial VAPPRO 837C at concentrations of 0.25, 0.50, 0.75, and 1.0 ml/100 ml (v/v) solution. The concentrated corrosion inhibitor CORPPRO was diluted in de-ionised water to reach the concentrations mentioned above for testing. To ensure the stability of the mixtures, the solutions were left to stand for one day before being used for testing.

Coating of metal sample VAPPRO 837C was applied on clean metal samples of carbon steel by immersion at three different times using a clock glass as a coating reservoir. Different immersion times were selected for the coating process (10 min, 20 min, 30 min, 8 h, and 24 h). After the coating, the samples were removed using tweezers and left to dry. Immediately after drying, the metal samples were tested using potentiostat to calculate their corrosion rates. Eventually, the immersion coating and drying time were also varied to establish the best condition for an optimum protection efficiency.

16.3.4 Elevated Temperature Drying

Figure 2 shows the set-up used for the elevated drying temperature test. The light bulb was first switched on to heat up the interior of the box. When the temperature inside the box stops fluctuating, metal samples that have just completed coating were clipped onto the string for drying. After drying, the samples were immediately tested using the potentiostat to determine their corrosion rates.

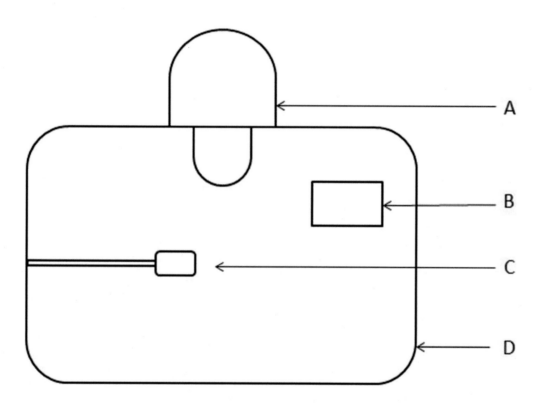

Figure 2. Elevated temperature drying set-up: (a) light bulb, (b) digital thermometer, (c) sample, and (d) container.

16.3.5 Freeze-Thaw Test

The freeze-thaw test was conducted following the ASTM D2243-95(2008) standard.[16] The formulations were first tested for their initial viscosity (before freezing). After the first testing, the formulations were stored at −18°C in a freezer for seventeen hours. Subsequently, the formulations were removed and thawed at room temperature for seven hours. Posterior to thawing, the formulations were placed back in the freezer at −18°C, and one more cycle of freeze-thaw was carried out. After the final thawing step was completed, the formulations were tested for their viscosity. The viscosity before and after freeze-thawing were compared to determine the freeze-thaw stability of the formulations.

16.3.6 FTIR Analysis

A Fourier Transform Infrared analysis was conducted on a 0.25 CORPPRO formulation using the attenuated total reflectance (ATR) method. A background scan was first carried out. After which, two to three drops of the formulation was added onto the zinc selenide crystal surface using a disposable dropper. The formulation was then scanned, and the resulting spectrum was analysed.

16.4 Results and Discussion

16.4.1 Polarisation Curves

Annex 1 shows the potentiodynamic polarisation curves (figures 4 to 7), and annex 2 presents the data from these curves. The results were grouped according to the drying time and compared with their concentrations. The behaviour of the polarisation curves depends on the inhibitor such as cathodic, anodic, and organic inhibitors.[18] For the organic inhibitors, the effect of the solution in the metal sample can be presented by a mixed behaviour.[19] The results in this work suggest that the behaviour of the polarisation curves could change because of different factors such as the concentration, the coating time, and the drying time. But it is possible to follow a tendency.

For all curves, the cathodic and anodic slopes showed the same behaviour, but the position of potential and current changed regarding the blank. With a drying time of 1 h, results showed that for any concentration, the polarisation curve had a nobler position when the coating time was below 30 min. As well, the value of the current decrease when a coating time was below 30 min. This behaviour is repeated on 2 h of drying time, but this changed when a 5 h or 24 h of drying time was applied. In this case, the nobler positions were obtained when a coating time was over 30 min.

As a result of this way of acting, samples with inhibitor presented an effect of corrosion inhibition. It was also observed in the values of corrosion efficiency and corrosion rate because all samples with inhibitor had a better performance in the environment. This may be due to the formation of a protective film of molecules on the metal surface and a mixed effect of inhibition. Different articles have suggested this mechanism of inhibition.[20, 21] Most VCIs form a film by surface adsorption upon hydrolysis or dissociation. The presence of amino groups, carboxylate anions, and the π bond between carbon atoms enhance the corrosion inhibiting.[22]

16.4.2 Corrosion Reduction Efficiency

The effect of drying time on corrosion reduction efficiency is shown in tables 6 to 9 (annex 2). The data gathered suggests that by increasing the drying time, the corrosion reduction efficiency becomes stable for any time of immersion or inhibitor concentration. It can be seen with a drying time of 5 h and 24 h where it presented with a small variation of efficiency for 10, 20, and 30 min for any concentration. However, the highest corrosion efficiency was provided with 1 h of drying time (91.29 per cent), but it is not uniform regarding the concentration.

This could be because while the coating appears to be dry visually after 20 min of drying, there could still be some water content present in the film, which promotes corrosion. It is also possible that the protective film of inhibitor molecules has not been properly organised for drying times before 5 h. This suggests that the optimum drying time should be around 24 h.

This optimum drying time of 24 h is also conducive for practical consideration of working hours in the industry. It would be easy to keep track of the end of the drying time and ensure that it ends within working hours so that the metals can be stored immediately after drying.

Similarly, the effect of concentration of CORPPRO can be interpreted. The highest corrosion reduction efficiency was provided with a CORPPRO concentration of 0.5 when the drying time was 1 h, 2 h, and 5 h. For 24 h, the highest corrosion reduction efficiency was obtained with a concentration of 0.25 (85.07 per cent); nevertheless, the concentration of 0.5 is close to this efficiency (82.27 per cent).

Through varying the coating time of CORPPRO, it was observed that the corrosion reduction efficiency of the formulations is different according to drying time. For 1 and 2 h of drying, the efficiency does not show a stable trend varying coating time. On the other hand, for 5 h and 24 h of drying, the corrosion reduction efficiency is stable from 10 min to 8 h of coating time. This suggests that increasing the coating time beyond 30 min has no significant effect on the corrosion reduction efficiency compared to a coating time of 30 min.

This could be because at 30 min, the surface of the metal substrate is fully saturated with inhibitor molecules and the metal surface-inhibitor interactions have stabilised. Therefore, any further coating time would not cause a decrease in corrosion, while a lower coating time would result in the substrate's surface not being saturated with inhibitor molecules, which correlates to a higher corrosion rate.

Considering the cost of time for a corporation, using a coating time of 10 min would be optimum as this would allow a fast-processing time without any loss in corrosion inhibition.

16.4.3 Effect of Drying at Elevated Temperatures

Through varying the temperature of drying, it was observed that drying at an elevated temperature of about 30°C obtains higher corrosion reduction efficiency. This can be seen from the 30°C drying obtaining a corrosion reduction efficiency of 58.6 per cent, while the 20°C drying obtained a corrosion reduction efficiency of 38 per cent, as shown in table 2 below.

Table 2. Effect of temperature during drying on corrosion reduction efficiency of 0.25 CORPPRO for 30 min coating time

Temperature (°C)	Corrosion reduction efficiency (%)
	0.25 CORPPRO 30 min coating 2 h room temperature drying
20	38
33	58.6

This could be because at elevated temperatures, the inhibitor molecules have more kinetic energy and are more mobile and thus more likely to orient themselves such that the polar functional groups face the metal surface to preferentially form a protective film over the surface of the metal substrate, therefore resulting in an increase in corrosion reduction efficiency for the metal sample that was dried at elevated temperatures.[17]

After 2 h drying at elevated temperatures, the surface of the metal sample still appeared wet on visual inspection. This is likely because the box used to contain the samples at elevated temperatures also limited the ventilation to a small area. This would reduce the rate of evaporation of the wet coating layer. Therefore, for future testing of drying at elevated temperatures, it would be suggested to carry out the drying in a well-ventilated area.

Compared to samples that were dried at room temperature and not contained in a box but left in a well-ventilated area and had a dry surface on visual inspection, the samples dried at elevated temperatures showed higher corrosion reduction efficiency. This suggests that the corrosion reduction efficiency is not directly related to the water content present in the coated film but rather on the complete formation of the protective film of inhibitor molecules.

16.4.4 Freeze-Thaw Tests

By varying the concentration of antifreeze (ethylene glycol) in the formulations, it was observed that all the formulations froze at −18°C. It was also observed that the viscosities of all the formulations remained around 29 cP after undergoing the freeze-thaw test, as shown in table 3. For all the formulations, there was no significant change in the viscosity after two cycles of freeze-thawing. Further testing is required to determine if the viscosity of the formulations will remain consistent after going through more freeze-thaw cycles.

The 0.25 CORPPRO formulations froze even after the addition of 1 per cent ethylene glycol antifreeze agent. To prevent the freezing of the formulations at −18°C, a higher concentration of antifreeze agent would have to be used.

Table 3. Effects of freeze-thawing on the viscosity of 0.25 CORPPRO formulations with and without 1 per cent antifreeze

% Antifreeze	Viscosity (cP)		
	Before freeze	After freeze	Freeze
Freeze thaw 1 with 1% antifreeze	28.2	30.9	Yes
Freeze thaw 2 with 1% antifreeze	29.3	28.3	Yes
Freeze thaw 1 w/o antifreeze	31.2	30.0	Yes
Freeze thaw 2 w/o antifreeze	28.0	30.6	Yes

16.4.5 pH test for varying concentrations of CORPPRO

Through varying the concentration of CORPPRO, it was observed that the pH is above 9 for all concentrations, as shown in table 4. The MSDS of CORPPRO states that it has a pH of 9, therefore it can be seen that diluting CORPPRO with distilled water does not affect its pH. For all formulations, the pH is below 11, and, thus, they will not cause skin irritation.

Table 4. Effect of concentration of CORPPRO on pH of formulations with different concentrations

CORPPRO concentration (vol%)	Average pH
0.25	9.60
0.50	9.60
0.75	10.60
1.0	10.7

Since the pH of the formulations does not vary much as the concentration of CORPPRO varies, it would not be viable to use pH as a measure to determine the concentration of CORPPRO formulations during manufacturing. It would be advised to use viscosity instead of pH to determine the concentration of formulations produced during manufacturing.

16.4.6 Functional Group Determination by FTIR

Figure 3 below shows the FTIR spectrum of 0.25 CORPPRO. The significant peaks were identified, and their correlated functional groups peaks are shown in table 5.

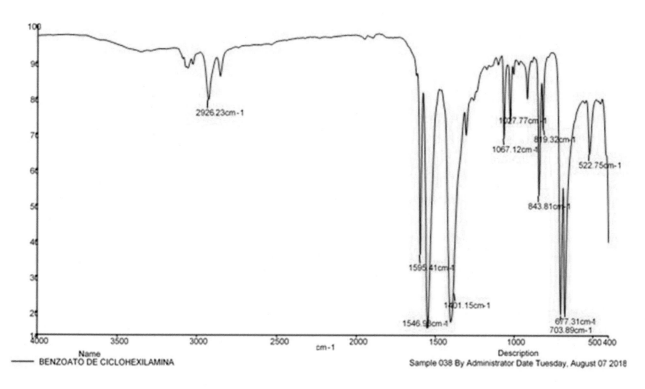

Figure 3. 0.25 CORPPRO FTIR spectrum.

Table 5. Correlated functional groups of significant peaks in 0.25 CORPPRO FTIR spectrum

Wavenumber (cm^{-1})	Functional group	Transition
≈3300	amine	–N–H stretch
3000–2850	alkanes	C –H stretch
1640–1550	primary and secondary amines	N–H bend
1599	carboxylate	COO– asym stretch
1390–1410	carboxylate	COO– sym stretch
1300–1450	aromatic	C –C stretch

The spectrum shown in figure 3 has a peak at around 2927 cm^{-1}. This could be due to carboxylate O–H stretching or alkane C–H stretching. The peaks at 1595 cm^{-1} and 1547 are due to COO$^-$ asymmetric stretch or primary/secondary amine N–H bending. The peak at 1401 cm^{-1} corresponds to carboxylate COO$^-$ symmetric stretching. The wavenumber, functional groups, and transition are presented in table 5.

Based on the absorption peaks present in the spectrum, it is likely CORPPRO contains an organic compound amine carboxylate aromatic type. This matches the description of CORPPRO in its MSDS, which states that CORPPRO contains 'amine carboxylate'.

16.5 Conclusions

The following conclusions can be derived based on the results obtained:
1. The optimum concentration for VAPPRO 837C formulations to achieve the highest corrosion reduction efficiency is 0.25 CORPPRO according to the optimum drying time.
2. The optimum processing parameters to achieve a short processing time without sacrificing corrosion reduction efficiency for coating carbon steel samples are 10 min coating and 24 h drying, as seen in effect of varying coating time on corrosion reduction efficiency and effect of varying drying time on corrosion reduction efficiency.
3. The formulations were found to have stable viscosity over a five-week period. The viscosities were also stable after freeze thawing for two cycles. This would suggest that there is no degradation of the VCI over the five-week period and from freezing. But further testing is required to determine the amount of antifreeze required to prevent freezing since the formulations with 1 per cent antifreeze still froze at −18°C. This can be seen in viscosity stability tests and freeze-thaw tests.

16.6 References

1. E. McCafferty, *Introduction to Corrosion Science*, Springer, 2010, 5.
2. M. G. Fontana, *Corros. Eng.*, McGraw-Hill, 1986, 1, 2.
3. E. Lyublinski, P. Lynch, I. Roytman, and T. Yakubovskaya, Application Experience and New Approaches for Volatile Corrosion Inhibitors, *Int. J. Corros. Scale Inhib.*, 2015, 4, no. 2, 176–192. DOI: 10.17675/2305-6894-2015-4-2-110-115.
4. N. N. Andreev and Y. I. Kuznetsov, Volatile Inhibitors of Metal Corrosion. I. Vaporization, *Int. J. Corros. Scale Inhib.*, 2012, 1, no. 1, 16–25. DOI: 10.17675/2305-6894-2012-1-1-016-025.
5. J. Granath, *Overview of Corrosion Protection with Volatile Corrosion Inhibitors*, ECS Transactions, 2010, 25, no. 30, 15–21. DOI: 10.1149/1.3321953.

6. M. Prenosil, *Volatile Corrosion Inhibitors*, Cortec Corp., 2001, 14–17.

7. National Research Council, *Research Opportunities in Corrosion Science and Engineering*, DC, The National Academies Press, 2011. DOI: 10.17226/13032.

8. N. N. Andreev, O. A. Goncharova, and S. S. Vesely, Volatile Inhibitors of Atmospheric Corrosion. IV. Evolution of Vapor-Phase Protection in the Light of Patent Literature, *Int. J. Corros. Scale Inhib.*, 2013, 2, no. 3, 162–193. DOI: 10.17675/2305-6894-2013-2-3-162-193.

9. I. A. Gedvillo, S. V. Oleinik, I. S. Sivokon, and N. N. Andreev, Laboratory Assessment of the Efficiency of Corrosion Inhibitors at Oilfield Pipelines of the West Siberia region V. Rotating cylinder and cage, *Int. J. Corros. Scale Inhib.*, 2013, 2, no. 4, 287–303. DOI: 10.17675/2305-6894-2013-2-4-287-303.

10. D. A. Bayliss and D. H. Deacon, *Treatment of the Air in Steelwork Corrosion Control*, CRC Press, 2003, 174.

11. J. A. Quitmeyer, Amine Carboxylates: Additives in Metalworking Fluids, *Soc. Tribol. Lubr. Eng.*, 1996, 52, 835.

12. L. H. Hihara, R. P. I. Adler, and R. M. Latanision, *Environmental Degradation of Advanced and Traditional Engineering Materials*, CRC Press, 2013, 98.

13. ASTM G-1(2003), *Standard practice for preparing, cleaning and evaluating corrosion test specimens*, West Conshohocken, PA, 2003, www.astm.org.

14. Singapore Drinking Water Quality (July 2017–June 2018), Singapore's National, Water Quality. https://www.pub.gov.sg/Documents/ Singapore_ Drinking_Water_Quality.pdf.

15. J. Holden, A. Hansen, A. Furman, R. Kharshan, and E. Austin, Vapor corrosion inhibitors in hydro-testing and long-term storage applications, NACE International, Conference paper No. 10405, Corrosion conference and Expo 2010.

16. ASTM D2243-95(2008), *Standard Test Method for Freeze-Thaw Resistance of Water-Borne Coatings*, ASTM International, West Conshohocken, PA, 2008, www.astm.org.

17. F. A. Ansari, C. Verma, Y. S. Siddiqui, E. E. Ebenso, and M. A. Quraishi, Volatile Corrosion Inhibitors for Ferrous and Non-ferrous metals and Alloys: a review, *Int. J. Corros. Scale Inhib.*, 2018, 7, no. 2, 126–150. DOI: 10.17675/2305-6894-2018-7-2-2.

18. B. E. Brycki, I. H. Kowalczyk, A. Szulc, O. Kaczerewska, and M. Pakiet, *Organic Corrosion Inhibitors*, IntechOpen, 2017.

19. G. Camila, D. Galio, and A. Galio, *Corrosion Inhibitors—Principles, Mechanisms and Applications*, IntechOpen, 2014.

20. L. M. Rymer, M. Sieber, S. Lautner, and F. Faßbender, *Operating Principle of Volatile Corrosion Inhibitors in the Jar Test*, IOP Conf. Series: Materials Science and Engineering, 2019.

21. M. Sieber, S. Lautner, and F. Faßbender, *Evaluation of Volatile Corrosion Inhibitors in the Presence of Condensation Water by Electrochemical Methods*, IOP Conf. Series: Materials Science and Engineering, 2019.

22. M. A. Quraishi and D. Jamal, *Development and Testing of All Organic Volatile Corrosion Inhibitors*, Corrosion, 2002, 58, no. 5, 387–391.

Annex 1

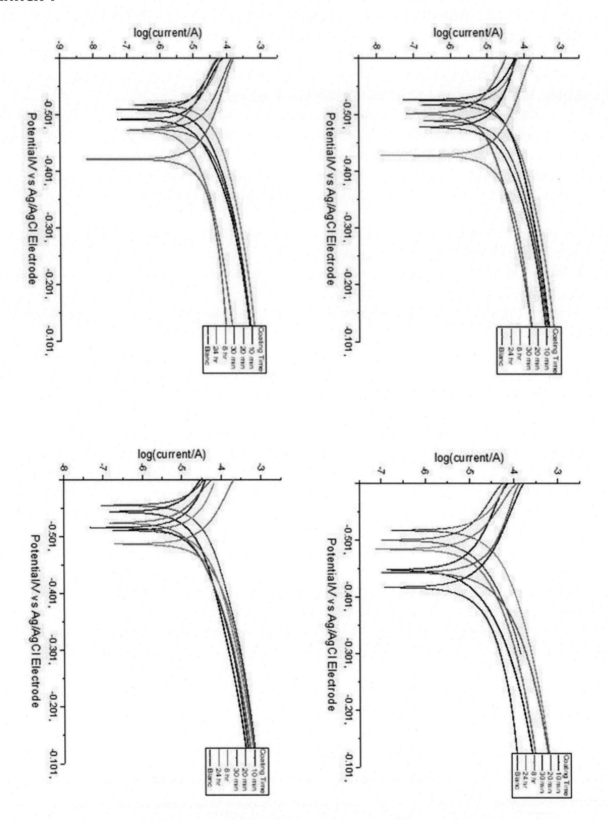

Annex 1 Figure 4. 1 h of drying. 0.25, 0.5, 0.75, 1.0 CORPPRO.

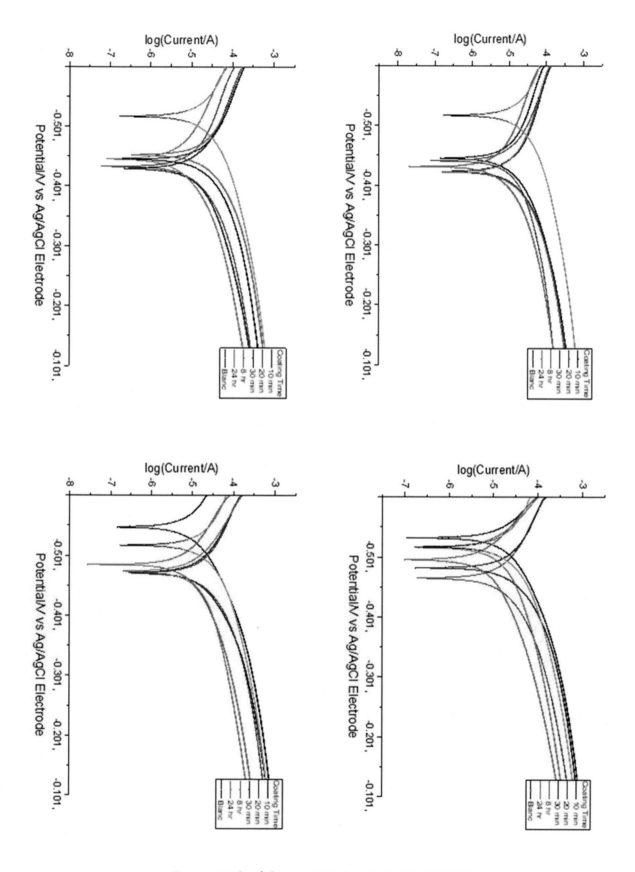

Figure 5. 2 h of drying. 0.25, 0.5, 0.75, 1.0 CORPPRO.

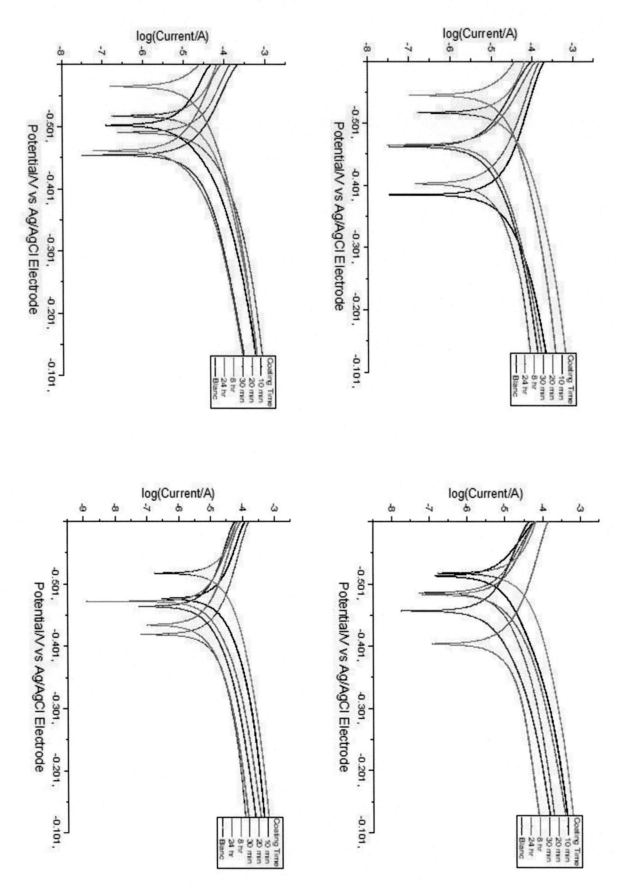

Figure 6. 5 h of drying. 0.25, 0.5, 0.75, 1.0 CORPPRO.

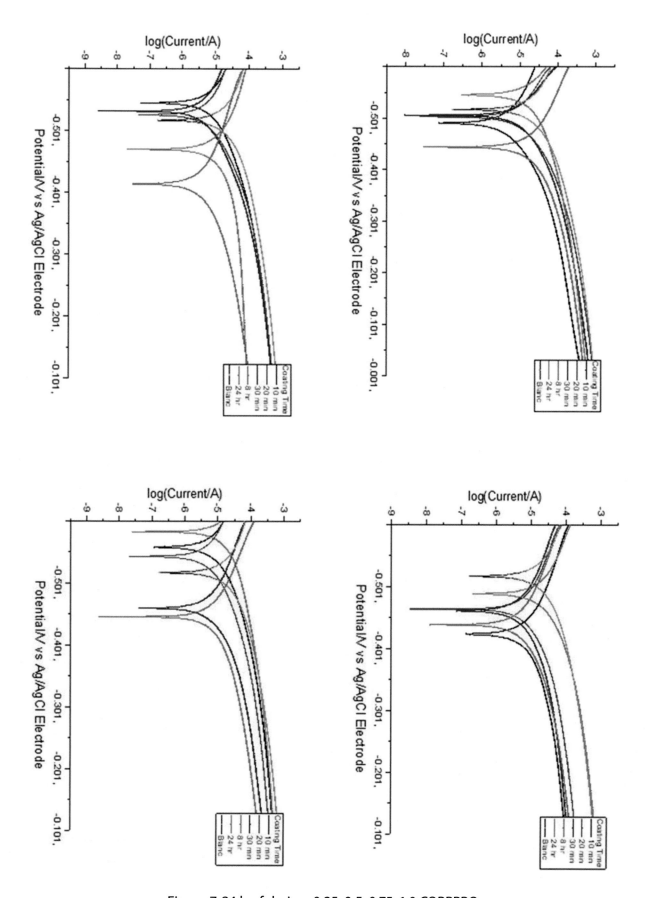

Figure 7. 24 h of drying. 0.25, 0.5, 0.75, 1.0 CORPPRO.

Annex 2

Table 6. Results obtained from polarisation curves for carbon steel with different concentrations of VAPPRO CORPPRO (drying time 1 h)

Concentration (vol %)	Coating time	Ecorr (V)	Icorr (μA cm⁻²)	Corrosion rate (mm/year)	Efficiency (%)
	Blank	−0.518	0.4576	0.643	
	10 min	−0.527	0.2893	0.501	22.08
	20 min	−0.49	0.2613	0.476	25.97
0.25	30 min	−0.479	0.2066	0.377	41.37
	8 h	−0.503	0.08183	0.12	81.34
	24 h	−0.429	0.2065	0.3815	40.67
	10 min	−0.545	0.1853	0.345	46.35
	20 min	−0.556	0.3094	0.571	11.20
0.50	30 min	−0.517	0.1971	0.364	43.39
	8 h	−0.525	0.03852	0.056	91.29
	24 h	−0.488	0.4198	0.463	27.99
	10 min	−0.449	0.1094	0.202	68.58
	20 min	−0.501	0.1115	0.426	33.75
0.75	30 min	−0.418	0.1831	0.338	47.43
	8 h	−0.443	0.3177	0.469	27.06
	24 h	−0.485	0.1445	0.266	58.63
	10 min	−0.492	0.1007	0.186	71.07
	20 min	−0.476	0.2449	0.453	29.55
1.0	30 min	−0.51	0.0347	0.064	90.05
	8 h	−0.422	0.1845	0.34	47.12
	24 h	−0.473	0.06921	0.102	84.14

Table 7. Results obtained from polarisation curves for carbon steel with different concentrations of VAPPRO CORPPRO (drying time 2 h)

Concentration (vol %)	Coating time	Ecorr (V)	Icorr (µA cm^{-2})	Corrosion rate (mm/year)	Efficiency (%)
0.25	Blank	−0.518	0.4576	0.643	
	10 min	−0.446	0.1655	0.305	54.25
	20 min	−0.442	0.1755	0.324	51.20
	30 min	−0.424	0.2111	0.39	40.61
	8 h	−0.432	0.08577	0.126	82.99
	24 h	−0.423	0.2035	0.3	55.06
0.50	10 min	−0.533	0.08052	0.148	79.45
	20 min	−0.466	0.2377	0.439	32.74
	30 min	−0.483	0.3701	0.583	9.63
	8 h	−0.497	0.08001	0.118	84.27
	24 h	−0.517	0.05817	0.107	86.04
0.75	10 min	−0.548	0.1441	0.212	69.18
	20 min	−0.471	0.2774	0.512	21.03
	30 min	−0.473	0.2236	0.413	36.92
	8 h	−0.485	0.09218	0.136	81.38
	24 h	−0.475	0.1726	0.319	52.01
1.0	10 min	−0.446	0.3358	0.618	4.01
	20 min	−0.434	0.1722	0.318	52.17%
	30 min	−0.43	0.2025	0.374	43.18%
	8 h	−0.453	0.3664	0.616	4.33%
	24 h	−0.477	0.07589	0.112	85.23%

Table 8. Results obtained from polarisation curves for carbon steel with different concentrations of VAPPRO CORPPRO (drying time 5 h)

Concentration (vol %)	Coating time	Ecorr (V)	Icorr (μA cm⁻²)	Corrosion rate (mm/year)	Efficiency (%)
	Blanc	−0.518	0.4576	0.643	
	10 min	−0.386	0.1999	0.37	42.46
	20 min	−0.404	0.1385	0.255	60.34
0.25	30 min	−0.463	0.07362	0.136	78.85
	8 h	−0.466	0.1022	0.15	76.67
	24 h	−0.546	0.2198	0.324	49.61
	10 min	−0.514	0.1928	0.356	44.63
	20 min	−0.487	0.07031	0.123	80.87
0.50	30 min	−0.458	0.04859	0.089	86.16
	8 h	−0.484	0.1034	0.152	76.36
	24 h	−0.405	0.1473	0.286	55.52
	10 min	−0.503	0.1051	0.194	69.83
	20 min	−0.492	0.3276	0.5651	12.12
0.75	30 min	−0.455	0.2146	0.396	38.41
	8 h	−0.462	0.1088	0.16	75.12
	24 h	−0.566	0.2143	0.316	50.86
	10 min	−0.477	0.2384	0.44	31.57
	20 min	−0.42	0.1613	0.298	53.65
1.0	30 min	−0.465	0.0885	0.164	74.49
	8 h	−0.473	0.0971	0.143	77.76
	24 h	−0.435	0.08684	0.128	80.09

Table 9. Results obtained from polarisation curves for carbon steel with different concentrations of VAPPRO CORPPRO (drying time 24 h)

Concentration (vol %)	Coating time	Ecorr (V)	Icorr ($\mu A\ cm^{-2}$)	Corrosion rate (mm/year)	Efficiency (%)
0.25	Blanc	−0.518	0.4576	0.643	
	10 min	−0.504	0.06517	0.096	85.07
	20 min	−0.507	0.1209	0.178	72.32
	30 min	−0.445	0.1174	0.173	73.09
	8 h	−0.547	0.2731	0.504	21.62
	24 h	−0.492	0.2689	0.397	38.26
0.5	10 min	−0.425	0.1191	0.175	72.78
	20 min	−0.465	0.1088	0.16	75.12
	30 min	−0.462	0.08547	0.126	80.40
	8 h	−0.44	0.0777	0.114	82.27
	24 h	−0.49	0.3139	0.463	27.99
0.75	10 min	−0.559	0.1274	0.188	70.76
	20 min	−0.544	0.09266	0.136	78.85
	30 min	−0.461	0.09375	0.138	78.54
	8 h	−0.447	0.1013	0.149	76.83
	24 h	−0.584	0.2314	0.341	46.97
1.0	10 min	−0.546	0.1061	0.156	75.74
	20 min	−0.527	0.0883	0.13	79.78
	30 min	−0.533	0.06786	0.1	84.45
	8 h	−0.415	0.05728	0.105	83.67
	24 h	−0.471	0.2035	0.3	53.34

CHAPTER

The Natural Gas Industry: Equipment, Materials, and Corrosion

Benjamin Valdez,[1] Michael Schorr,[1] and Jose M. Bastidas[2]

[1]Laboratory of Materials, Minerals and Corrosion, Institute of Engineering, CP 21280, Mexicali, Mexico,
[2]National Centre for Metallurgical Research, Spanish Research Council (CSIC), 28040 Madrid, Spain

17.1 Abstract

Corrosion is a crucial worldwide problem that strongly affects the oil and gas industry. Natural gas (NG) is a source of energy used in industrial, residential, commercial, and electric applications. The abundance of NG in many countries augurs a profitable situation for the vast energy industry. NG is considered friendlier to the environment and has lesser greenhouse gas emissions compared with other fossil fuels. In the last years, shale gas is increasingly exploited in the USA and in Europe, using a hydraulic fracturing (fracking) technique for releasing gas from the bedrock by injection of saline water, acidic chemicals, and sand to the wells. Various critical sectors of the NG industry infrastructure suffer from several types of corrosion: steel casings of production wells and their drilling equipment, gas-conveying pipelines including pumps and valves, plants for regasification of liquefied NG, and municipal networks of NG distribution to the consumers. Practical technologies that minimise or prevent corrosion include selection of corrosion-resistant engineering materials, cathodic protection, use of corrosion inhibitors, and application of external and internal paints, coatings, and linings. Typical cases of corrosion management in the NG industry are presented based on the authors' experience and knowledge.

Keywords: cathodic protection, corrosion, corrosion inhibitors, natural gas, pipelines

17.2 Introduction

Environmental quality, clean energy, and worldwide water scarcity have been established today as central disciplines in modern science, engineering, and technology. They are linked to the crucial problems of climate change and global warming (Raichev, Veleva, and Valdez 2009; Roberge 2009; Valdez et al. 2012).

Today, it is generally accepted that corrosion and pollution are interrelated harmful processes because many pervasive pollutants accelerate corrosion, and corrosion products such as rust, salts, and oxides also pollute water bodies. Both are pernicious processes that impair the quality of the environment, the efficiency of the industry, and the durability of the infrastructure assets. To complicate the situation, some contaminated water bodies also serve as sources of potable water (Marcos, Botana, Valdez, and Schorr 2006; Schorr and Valdez 2006, 2007; Schorr, Valdez, and Quintero 2006).

Recent damaging cases of joint corrosion and pollution in which H_2S was a prominent factor have been reported (NACE 2009; Rebak 2001). This situation is aggravated when industrial, municipal, and agricultural pollutants are discharged into adjacent water bodies such as lakes, rivers, and estuaries (Schorr and Valdez 2005).

In power-generating plants that burn fossil fuels such as coal, oil, and natural gas (NG), the last is generally preferred because of its transportation and production efficiency and favourable quantitative carbon/hydrogen ratio, and it produces less combustion pollutants.

The April 2011 earthquake and tsunami in Fukushima, Japan, damaged electricity-generating nuclear plants, which led to problems of national security, industrial safety, and hazardous radiation and to a fundamental examination of the future of the nuclear industry by many countries (Park 2011). Recently, workers at the wrecked Fukushima plant were exposed to radiation after contaminated water leaked from several pipes. The German government panicked into ordering the closure of several nuclear plants. Renewable energies such as solar and wind are intermittent, surging with the weather. Following these changes, the European aerial electricity grid, with its steel transmission towers and insulated aluminium and copper cables, is being expanded and modernised to supply the energy required by the increasing population. As a consequence of this critical situation, purified NG is now considered by European and American energy authorities as a 'green' alternative, with lower corrosion and no radioactive or toxic pollutants (So, Valdez, and Schorr 2013).

One of the latest developments in the struggle against corrosion is the creation of a central institution to serve the US Armed Forces in dealing with corrosion of military facilities, equipment, and weapons, which is one of the major consumers of energy, including NG (Dunmire, Thomson, and Yunovich 2005; Greenwood 2013; Hummel 2014). Thanks to better NG technology and improved efficiency, it is becoming cleaner and more plentiful.

The economic and social relevance of the NG industry is evident in the activities of diverse international and national professional associations, R & D institutions, and industrial enterprises involved in all aspects of NG science, engineering, and technology (table 1). These include authorities from the government, industry, and academia who are addressing the progress in NG industry and its vital importance in national and global prosperity. Public and commercial journals publishing useful information on the NG industry are listed too.

Table 1. Associations, organisations, and enterprises dealing with NG engineering and technology

Associations and organisations	Website
American Gas Association	www.aga.org
American Petroleum Institute	www.api.org
American Public Gas Association	www.apga.org
ASM International	www.asminternational.org
Asociación Mexicana de Infraestructura Portuaria, Marítima y Costera, AC	www.amip.org.mx
ASTM International	www.astm.org
Energy Information Administration, USA	www.eia.gov
Energy Institute, London, UK	www.energyinst.org
Interstate Natural Gas Association of America	www.ingaa.org
Israeli Institute of Energy and Environment	www.energy.org.il
NACE International	www.nace.org
Natural Gas Authority	www.energy.gov.il
Natural Gas Europe	www.naturalgaseurope.com
Petróleos Mexicanos	www.pemex.com
Pipeline Research Council International	www.prci.org
The Israel Chemical Society	www.chemistry.org.il
US Department of Energy	www.energy.gov

Journals	Website
Advanced Energy Materials	onlinelibrary.wiley.com
Energy	www.journals.elsevier.com
Energy Materials	www.maneyonline.com
Environment International	www.journals.elsevier.com
Environmental Pollution	www.journals.elsevier.com
Hydrocarbon Processing	www.gulfpub.com
Journal of Materials Chemistry	pubs.rsc.org
Journal Natural Gas Science and Engineering	www.journals.elsevier.com
Materials Performance, NACE	www.nace.org
Mexico Oil and Gas Review	www.mexicooilandgasreview.com
Natural Gas & Electricity	onlinelibrary.wiley.com
Offshore Magazine	www.offshore-mag.com
Oil and Gas Journal	www.ogj.com
Oil and Gas Facilities Magazine	www.spe.org
Journal of Petroleum and Gas Engineering	www.academicjournals.org
Pipelines International	www.pipelinesinternational.com
World Oil	www.worldoil.com

17.3 Natural Gas

NG is a source of energy for industrial, residential, commercial, and electric applications. It is also employed as a raw material for the production of polymers and plastics. The share of NG in the total energy consumption in the USA in 2013 was 23.7 per cent (Lucas 2015). The main sectors of the NG industry include drilling, production, storage, transportation, liquefaction, vaporisation, transmission, and distribution, and all are affected to some extent by corrosion. NG is obtained from petroleum offshore wells using drilling and production marine platforms. It is transported to the marine coast through submarine pipelines. NG is obtained from onshore wells too. Usually, it is extracted with salty or briny water and corrosive gases, mainly H_2S and CO_2 (Mokhatab, William, and James 2006; Smith 1990; Wang 2009). The principal component is CH_4, but it contains other light hydrocarbons as well (table 2). H_2S present in NG oil wells generates sour gas and corrosion problems, denoted as sour corrosion (Quintero Nuñez, Valdez, and Schorr 2010; Rebak 2001; Roberge 2009; Valdez and Schorr 2011; Valdez, Schorr, So, and Eliezer 2011; Wang and Atrens 2004). It is a weak, reducing water-soluble acid that can lead to the pitting of steels. An FeS film forms on steel surface, which is cathodic to the steel, enhancing further corrosion.

NG also contains CO_2, which converts into carbonic acid (H_2CO_3) when it is exposed to moisture, provoking corrosion and sometimes forming carbonate scales in well casings (Chilingar, Mourhach, and Al-Qahtani 2008; Heidersbach 2011; Philippine Corrosion Society 2009; Waard, Lotz, and Milliams 1991).

Wet acid gas, composed of a mixture of H_2S, CO_2, and water separated from offshore oil wells, is strongly corrosive to carbon steel (CS). This acidic gas is treated with amines and then compressed and injected into wells.

NG is treated to remove impurities and corrosive components, which are removed by applying membranes that prevent corrosion (Baker and Lokhandwala 2008). It is burned in power stations to generate electric energy.

Liquefied natural gas (LNG) is transported from continent to continent in cryogenic sea vessels (figure 1) to marine ports and regasification plants (RPs). These giant steel and aluminium (for the sailors' living quarters) ships are able to transport LNG from Indonesia to the Pacific Coast, from Algeria to the great cities of the Atlantic Ocean littoral and the Gulf of Mexico, and from NG countries on the Persian Gulf to the Mediterranean and the Baltic Sea coasts. Therefore, large quantities of NG are produced in countries where production far exceeds demand. It is converted to LNG by cooling and compressing after purification to remove water, solid particles, acidic gases, and heavy hydrocarbons.

Table 2. NG composition

Name	Formula	Volume (%)
Methane	CH_4	>85
Ethane	C_2H_6	3–8
Propane	C_3H_8	1–2
Butane	C_4H_{10}	<1
Pentane	C_5H_{12}	<1
Carbon dioxide	CO_2	1–2
Hydrogen sulphide	H_2S	<1
Nitrogen	N_2	1–5
Helium	He	<0.5

Figure 1. Cryogenic vessel carrier.

17.4 Shale Gas

Shale gas (SG) is a type of NG that is entrenched in rock formations constituted mainly by layers of shale, a silicate mineral. It has been known in the USA since the early twentieth century, but it has only been used as a central source of energy in the last decade. The US Department of Energy believes that the use of SG will dramatically reduce greenhouse emissions. Europe's SG deposits almost match those across the Atlantic; many countries might enjoy a bonanza of cheap gas if production, corrosion, and environmental problems are overcome (Anonymous 2013, 2015). The production of SG is increased by a mechanochemical technique of hydraulic fracturing also called fracking, which is done by injecting large quantities of salty water, inorganic acids (HCl, H_2SO_4), and sand under pressure into the SG wells to break up the rock bed and increase the SG flow. This aggressive mixture of acid and saline water may corrode the gas wells' steel equipment such as tubings, casings, pumps, and valves.

Some ecologists and environmentalists claim that fracking pollutes the air and drinking groundwater, but the evidence presented by industry cycles suggests that any such pollution is limited (Kargbo, Wilhelm, and Campbell 2010). Because of the SG boom, many coastal RPs are being closed because there is no need to import LNG. Meanwhile, companies seeking to exploit SG formations are first required to conduct environmental impact studies.

17.5 Natural Gas Industry

17.5.1 Infrastructure

The petroleum and NG industries are two key sectors of energy infrastructure, a powerful index of the vitality of a nation. NG is one of the most abundant sources of energy available today, and with continued innovation, it could provide cleaner energy for future generations. In many regions of the world, deals are in the work among countries, covering exploitation, production, and transportation of both oil and NG. Countries with no access to the sea are planning and installing long land steel pipelines for the exportation of NG, following the increasing global demand of energy.

The main sectors of the NG industry, arranged in chronological sequence of operation, are well drilling, production, cleaning, storage, transportation, liquefaction, vaporisation, transformation, and distribution.

For the sake of brevity, this review deals with corrosion problems of the central assets of the oil and gas industry infrastructure, as follows:

- drilling and exploitation of NG onshore wells
- marine petroleum platforms, their service vessels, and submarine pipelines
- pipelines for conveyance and distribution of NG
- LNG RPs

The corrosive characteristics of NG require the selection of corrosion-resistant alloys that will ensure long service life without corrosion. They include martensitic, austenitic, and duplex stainless steel (SS), precipitation-hardened steel, and acid-resistant Ni alloys. To maintain continuous and effective operation, a diversified assembly of equipment is employed in the NG infrastructure. These various equipment and their corrosion-resistant engineering materials are listed in table 3.

Table 3. NG industry: equipment and materials

Equipment	Materials
well casing	API steels
drilling tools	API steels
marine platforms	steels, reinforced concrete, SS
submarine pipelines	SS
port installations	steel, reinforced concrete
LNG pipes and pumps	austenitic SS, UNS S31600
LNG storage tanks	reinforced concrete lined with Ni-steel alloy
LNG vaporiser	aluminium alloy UNS A95052
seawater pipes	austenitic SS, UNS S31600
pumps and valves	ferritic, austenitic SS, Duplex S32250
steam and gas turbines	SS, Ni alloys
water and chemical storage tanks	fibreglass-reinforced plastics, epoxy, and polyester
heat exchangers: shell and tube, plates	Steel, aluminium, copper, SS, Ni alloys
NG pipelines	API steels, 5L
control instrumentation	metals, plastics

17.5.2 Production Wells

The well is the dominant component in oil and gas production. Drilling and extraction are the main operations where the corrosion problems arise because of the critical downhole corrosive factors: pressure of H_2S and CO_2 gases, briny water containing ionic-dissociated mineral salts (e.g., NaCl, $MgCl_2$), and elevated temperature (Beavers and Thompson 2006; NACE 2009; Wong-Dickason and Thomson 2013).

The well-drilling strings are made of high-strength, low-alloy CS; they fail through corrosion fatigue and environmental cracking (Cabrini, Lorenzi, Marcassoli, and Pastore 2011). Drilling muds facilitate the operation; their pH is maintained in the 8.5–11.0 range to control steel corrosion.

The well is separated from the downhole environment by a steel casing that is cemented to the rock formation. A central producing steel tube is used for extracting the mixture of oil, gas, and water to the oilfield surface for separation, cleaning, storage, and transmission. The internal surface of the casing and the exterior surface of the production tube suffer from corrosion, which is minimised by internal coatings and corrosion inhibitors. In addition, mineral scales are formed on the casing and tube surface (e.g., $CaCO_3$, $CaSO_4$, silicates), which provoke localised corrosion under deposits.

17.5.3 Marine Petroleum Platforms

An offshore platform is a complex structure that is used to drill wells, extract and process oil and NG, and deliver them onshore. Fixed and floating platforms are built using two basic engineering materials: steel and reinforced concrete. They are constructed at European and Asian shipyards, towed out to sea, and installed at various depths in seas around the world, usually near the coast. The platforms are served by oil storage ships and other vessels and helicopters for transportation of personnel and supply of equipment.

The platforms are provided with a flare stack made of corrosion-resistant Ni alloys to burn the excess unused gas (figure 2).

The sea is a dynamic system in permanent motion, with complex surface currents and winds flowing over its surface, generating waves that affect marine installations and vessels. The central corrosive factors are salinity, dissolved oxygen, pH, velocity, and temperature (Heidersbach, Dexter, Griffin, and Montemarano 1987). Corrosion is a major problem with offshore platforms because of the harsh environment; it appears at three levels: the atmosphere zone with salt-laden aerosols and sun radiation; the splash/tidal zone with violent waves and whitish oxygenated foam; and the submerged and muddy zone, which sometimes contain corrosive pollutants.

Figure 2. Semi-submersible petroleum platform fitted with a gas flare tower.

Floating marine platforms stand in 500 m shallow waters, but water elsewhere reaches depths of 2300 m and beyond. There are four thousand fixed platforms off the US coast and three hundred in the Bay of Campeche. Eighty per cent of Mexico's petroleum and 30 per cent of its NG are extracted in this area. The nearby maritime oil terminals have facilities to control and threat oil spills during loading operations into huge tankers bound toward USA and Europe.

Sea microorganisms and microorganisms settle down and adhere on installation surfaces and ships' hull, enhancing corrosion (Acuña, Schorr, and Hernandez 2004; Dürr and Jeremy 2009). Special paints, charged with biocides, may prevent or avoid these pernicious phenomena. Leakage of petroleum, which contains CO_2 and H_2S, from wells and submarine steel pipelines, results in corrosion damage and environmental pollution.

17.5.4 Natural Gas Pipelines

Pipes manufactured from CS play an important role in the global economy; they are used for the transportation of numerous fluids: potable water, municipal sewage, petroleum and its derivatives, hydrogen, iron ore slurry, sand oil, and, particularly, for the conveyance of NG (Haeseldonckx and D'haeseleer 2007). These pipelines operate mostly underground but are detected by facilities and auxiliaries on the ground such as pumps, huge valves, and equipment for flow control and cathodic protection (CP) (ANSI 2007; NACE 1998; Srinivassan and Eden 2006).

In the USA, there are half a million kilometres of NG pipelines with diverse diameters (up to 1.2 m) and wall thickness between 5 and 7 mm, depending on the transmission rate, the corrosiveness of the NG being transported, and the aggressiveness of the soil (Hanafy 2015). Pipelines suffer severe corrosion events, leading to leakage of NG and subsequent explosions and fires, with loss of human life and property. Pipeline network faces closures after leaks and deadly events, where even the slightest uncontrolled increase in temperature could lead to costly accidents. A quantitative method for risk assessment in NG pipelines is applied during the planning and building stages of a new pipeline, including urban NG pipeline network and modification of a buried pipeline, since it improves the level of safety (Han and Weng 2011; Hernandez et al. 2007; Jo and Ahn 2005).

CS, particularly those specified by the American Petroleum Institute and other European regulation agencies, are preferred for NG pipelines. API promotes the use of type 5D for drill pipe, 5CT for casing and tube, and 5L for pipeline. In some cases that require enhanced mechanical properties, low-alloy steel containing minor amounts of Mn, Ni, and Cr are applied. Different internal and external cases of corrosion occur in pipelines, such as corrosive wear, pitting, galvanic, microorganism-induced corrosion, erosion-corrosion, and stress (Bullard et al. 2002; Hernandez et al. 2007; Wang 2009). International organisations such as NACE, ASTM, ASM, and ISO have devoted much effort to combat and mitigate corrosion in NG pipelines, preparing and publishing books, handbooks, and standards for corrosion control.

17.5.5 LNG Regasification Plants

In 2012, around 100 RPs were operated worldwide, including plants in Mexico, Latin America, the USA, Europe, and Asia. There are a number of technologies for exporting NG energy from oil and gas fields to market, including pipelines, LNG, compressed NG, gas to wire (i.e., electricity) (Sidney and Dawe 2003). Various installations, structures, and equipment are utilised in RPs to convert LNG back to its gas form. They comprise two central units: the port terminal for the cryogenic ships' mooring and the open-rack vaporisation for conversion, which is heated by seawater, since the LNG temperature is −160°C (figure 3). After regasification, the NG is transferred to an onshore pipeline distribution system for the ultimate consumers (Quintero et al. 2014; Valdez, Schorr, Quintero, García, and Rosas 2010).

The RP equipment is fabricated from a wide spectrum of engineering materials, metallic and non-metallic, that display reasonable endurance to fluids (liquid, vapour, and gases) handled and processed in the RP installations and facilities. The main LNG equipment and their engineering materials are listed in table 3. Onshore installations and equipment include the chemical facilities used in the RP operation and service for the production of chlorine, alkali, ammonia, nitrogen gas, potable water, electricity, etc. Figure 4 presents a block diagram of an RP.

Figure 3. LNG rack vaporisers.

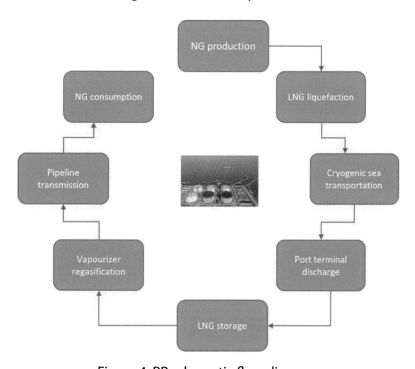

Figure 4. RP schematic flow diagram.

Because of the locations of the RPs on ocean and sea coasts and because of the complex equipment being operated, several types of corrosion are distinguished:

- marine corrosion, particularly in splash zones where violent waves break down and generate whitish oxygenated foam, which increases local corrosion
- Coastal atmospheric corrosion with salt-laden mist that condenses on the equipment surface during cold nights. Relevant environmental parameters are air temperature, relative humidity

(sometimes <70 per cent), salinity deposition rates, rainfall, wind effect, and air pollution, such as SO_x and NO_x from fossil fuel combustion, which causes acidic rain and subsequent corrosion.
- industrial corrosion of land equipment and machinery by humidity, dust deposition, galvanic couples, and surface paint deterioration.

Corrosion has a major impact on the economics of LNG facilities, safety, and environmental preservation.

17.6 Corrosion and Protection Control

The concepts of inspection to determine the conditions of a system, monitoring to assess the need for corrosion control in laboratory or field/plant testing, and evaluating the equipment and materials are essential to ensure the efficient operation of the NG facilities. A special technique was developed and applied for the detection of localised and general corrosion of NG transmission of pipelines using electrochemical noise sensors (Bullard et al. 2002). An additional technology employs sensors compatible with a robotic vehicle for in-line inspection that can manoeuvre within the pipe (Bickerstaff, Vaughn, Stoker, Hassard, and Garrett 2002).

Practical methods that minimise or eliminate corrosion include the selection of suitable corrosion-resistant engineering materials and application of paints and coatings to CS equipment, CP mainly for steel pipelines, and corrosion inhibitors for NG. Frequently, corrosion, scaling, and fouling phenomena appear simultaneously in vast and diverse installations and equipment of the NG industry. NG contains some associated water, which becomes acidic because of the presence of dissolved gases, mainly H_2S and CO_2. Scaling and fouling often derive from the salty water generated together with oil and gas in the production wells. Dissolved CO_2, forming carbonic acid, may react with Ca^{2+} salts, depositing $CaCO_3$ scale in CS surfaces.

17.6.1 Materials Selection

Materials selection is the dominant way to avoid corrosion failures in the different sectors and equipment of the NG industry (table 3). The technical process of selection comprises two main stages: a thorough analysis of the chemical media and the mechanical operating condition of the NG facility. The final choice is based on durability, cost, availability, ease of fabrication and repair, maintenance, and safety. This information is essential for making decisions about the type, urgency, and expenses in preventive and curative measures to be taken.

Most equipment is fabricated from CS in accordance with API 5L or additional API steels (see section 4.3), but equipment in contact with sour gas or seawater are constructed from martensitic and/or austenitic SS (Bhat 2011; Javaherdashti, Nwaoha, and Tan 2013).

17.6.2 Paints and Coatings

Paints and coatings have proprietary technology that changes with time, and their quality depends on local producers and suppliers. It is the standard process used in controlling external corrosion of offshore structures, pipelines, and process vessels. The paint film and solid coating should be impermeable to NG, moisture, and oxygen and have good adherence. The most important step in their application is correct surface preparation, following the producer instructions. The integrity and functionality of the pipelines is promoted by enterprises that develop and produce special paints and coatings for steel pipelines (Papavinasam and Revie 2008; Papavinasam, Attard, and Revie 2008).

17.6.3 Cathodic Protection

The combination of coating as the primary method of corrosion control, with CP as a supplementary method, is the most economical way to control corrosion—in particular, for submerged steel structures and pipelines. CP is applied by two techniques: sacrificial anodes (figures 5 and 6), made of Al, Mg, and Zn; and an impressed direct current at an electrochemical potential sufficiently negative to convert the steel structure in a cathodic surface. CP efficiency depends on the electrical conductivity of the environment and the continuous protection function of the sacrificial anodes (Garcia et al. 2015).

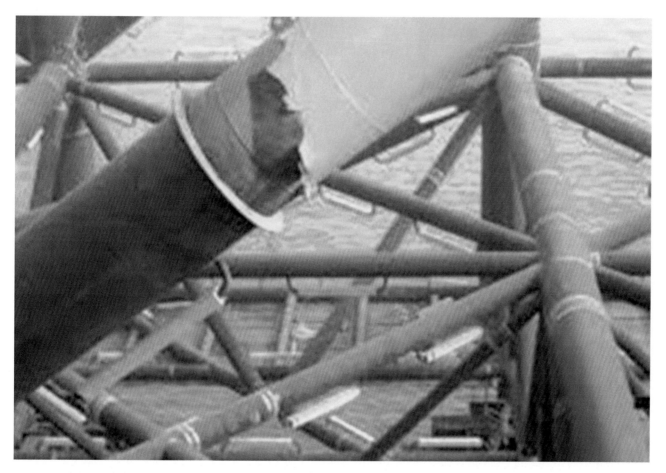

Figure 5. Legs of a petroleum marine platform fitted with aluminium alloy sacrificial anodes.

Figure 6. Ship propeller and rudder with aluminium alloy sacrificial anodes.

17.6.4 Corrosion Inhibitors

Corrosion inhibitors decrease the rate of attack by NG corrosive components. According to their corrosion control mechanism, they are classified as anodic oxidation inhibitors, cathodic inhibitors, film-forming inhibitors, scavengers for annihilation of H_2S or oxygen, and vapour-phase inhibitors. Some multicomponent inhibitors are often mixtures of those inhibitors, depending on the characteristics of the NG system, its water content, and operating temperature (Garcia et al. 2015; Inzunza, Valdez, and Schorr 2013; Sastri 2012).

Corrosion resistance is the main property that must be considered in the choice of materials for the NG industry equipment, but the final selection must be a compromise between technological and economic factors. Sometimes, it is more economical to use a higher-priced material than a lower-priced material that may require frequent maintenance or replacement.

17.7 Discussion

NG is an abundant resource across the USA. New extraction methods have led to a dramatic change in its supply. Years ago, virtually all NG production came from traditional gas or wells reached by vertical drills. In 2009, according to the Energy Information Administration, about 20 per cent of American NG came from shale formations. Compressed NG can be used for motor vehicles, after some adaptation.

The SG boom is increasing American NG exports, but it is closing coastal RPs that imported LNG from countries far away. NG and oil companies work to secure critical infrastructure against cyber threats and terrorist attacks (Melman 2014).

On the one hand, decreasing oil prices harm Mexico's economy. On the other hand, Mexico is undergoing an intensive reform process of the energy sector, including its oil, NG, and electricity industries. Mexico's abundant resources such as deep-sea oil in the Gulf of Mexico (Hernandez, Schorr, Carpio, and Martinez 1995; Raichev et al. 2009; Veleva, Castro, Hernandez-Duque, and Schorr 1998) and SG will be exploited, and additional refineries and pipelines are being planned and built with the active participation of heavy foreign investments. The reform was recently approved by the Mexican Parliament (Anonymous 2013; Mexican Energy Reform 2013) and is setting Petróleos Mexicanos (PEMEX), the national oil company, on the way to becoming a world oil enterprise, producing 3 million bpd (477 million L/day) (Duran 2014; Garcia et al. 2015).

The oil and NG industries in Mexico, including LNG RPs and the profitable petrochemical industry, are managed by PEMEX. These industries deal with exploration, production, transportation, and distribution of crude oil and its many derivatives.

To maintain a continuous and efficient operation, a massive assembly of structures and equipment is used, including marine petroleum platforms, service vessels, and submarine pipelines; semi-submersible vessels for drilling offshore petroleum wells; land pipelines for conveying and distributing; fuel storage tanks; sea wind turbines; oil transportation tankers, cargo ships, and harbour structures; heat exchangers; oil well casings; etc. All these diversified facilities and installations require the applications of protection with sacrificial anodes, which are produced by local Mexican companies (Garcia et al. 2015).

Recent discoveries of new sources of NG along the coast of the Mediterranean Sea have altered the energy balance of the region. The use of NG started in Israel in 2004, mainly by the Israel Electric Corporation and supported by the Natural Gas Authority (2015) (table 1) in the Ministry of Energy and Water Resources. NG is extracted from wells operated by marine platforms and passes a system of steel pipes, until it reaches reception stations on the country shore. The gas flows through a nation transmission system to supply the country and other neighbours. The development of NG fields by Israel, Syria, and Lebanon is paving the way toward a new reality in the region—of a probable useful coordination about NG exploration security (Melman 2014). Israel proposes to build a 'Euro' pipeline, in cooperation with Cyprus, Greece, and Italy, to supply NG to Europe.

The Technion-Israel Institute of Technology has established a postgraduate MSc degree program in energy source engineering and technology. The Israel Chemical Society organises symposia and courses on the application of NG in chemical, electricity, water, and transportation industries, including corrosion control.

Europe's SG deposits almost match those across the Atlantic in America. Determining which countries might enjoy cheap NG is not yet certain. Many things are in flux, including extraction technologies and production rates. Meanwhile, the European Union (EU) is being supplied by a long pipeline system from Russia, passing through Ukraine. The EU NG network has become more integrated with the installation of interconnector pipes. Countries in Central Europe can receive NG from the west via Germany, from Russia through alternative pipelines such as the Nord Stream, or from North African countries through submarine pipelines installed on the Mediterranean seabed. European national corrosion institutes and experts oversee and manage the corrosion problems of this huge system of pipes, pump stations, and valves to regulate NG flow and supply (figure 7).

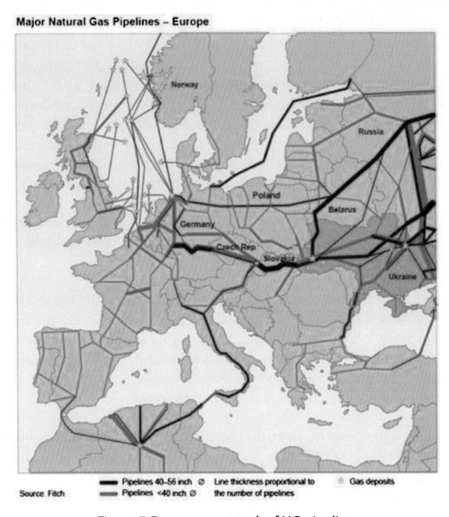

Figure 7. European network of NG pipeline.

17.8 Conclusions

- NG is the preferred fossil fuel for the operation of electricity-generating plants because of its low cost, ease of transport, and low pollution.
- The world's abundant SG beds in many countries in Europe and in America are becoming the most important factor in the global energy revolution.
- The production well is the main component of the oil and gas system. Their steel drilling strings, well casing, and tubing are attacked by downhole corrosive factors.
- The NG infrastructure requires the implementation of corrosion control methods through the stages of design, construction, installation, and operation.
- Corrosion prevention and mitigation is achieved by the application of modern corrosion control technology including the correct selection of engineering materials, covering the surface of steel equipment by chemical conversion coating and industrial paints, employing CP with sacrificial anodes for marine platforms and impressed current for pipelines, and choosing suitable corrosion inhibitors.

17.9 References

1. Acuña N., Schorr M., Hernandez G. Effects of marine biofilms on the fatigue resistance of an austenitic stainless steel. Corros Rev 2004; 22: 101–114.

2. Anonymous. Shale gas in Poland. Available at: www.economist.com/2012/shale-gas-poland. Accessed 2013.

3. Anonymous. Science and Technology Section. The Economist, February 2 and 16, 32 and 53, 2015.

4. ANSI/NACE SP607-2007/ISO 155-89 (MOD). Petroleum and natural gas industries—cathodic protection of pipeline. Houston, TX: NACE, 2007.

5. Baker R. W., Lokhandwala K. Natural gas processing with membranes: an overview. Ind Eng Chem Res 2008; 47: 2109–2121.

6. Beavers J. A., Thompson N. G. External corrosion of oil and natural gas pipelines. In: ASM handbook, vol. 13C. Corrosion: environments and industries. Materials Park, OH: ASM International 2006: 1015–1025.

7. Bhat S. Construction materials for acid gas and injection wells. Mater Perform 2011; 5: 62–66.

8. Bickerstaff R., Vaughn M., Stoker G., Hassard M., Garrett M. Review of sensor technologies for in-line inspection of natural gas pipelines. Albuquerque, NM: Sandia National Laboratories, 2002.

9. Bullard S. J., Covino Jr. B. S., Russell J. H., Holcomb G. R., Cramer S. D., Ziomek-Moroz M., Albany O. R. Electrochemical noise sensors for detection of localized and general corrosion of natural gas transmission pipelines. Final report, Department of Energy, contract FE-01-06, 2002.

10. Cabrini M., Lorenzi S., Marcassoli P., Pastore T. Hydrogen embrittlement behaviour of HSLA line pipe steel under cathodic protection. Corros Rev 2011; 29: 261–274.

11. Chilingar G. V, Mourhach R., Al-Qahtani G. D. The fundamentals of corrosion and scaling for petroleum and environmental engineers. Gulf Publishing, 2008.

12. Dunmire D., Thomson N., Yunovich M. Cost of corrosion and corrosion maintenance strategies. Corros Rev 2005; 25: 247–261.

13. Duran J. A. Effects of alloys and pollutants on the behaviour of aluminum anodes for cathodic protection. MSc thesis, University of Baja California, 2014.

14. Dürr S., Jeremy T. Biofouling. New York: Wiley, 2009.

15. Garcia I., Valdez B., Schorr M., Herrera T., Duran J., Ramirez V. Sacrificial anodes for cathodic protection in the oil industry in Mexico. Mater Perform 2015; 54: 38–42.

16. Greenwood C. Corrosion Office issue report on the impact of corrosion on the military facilities. Summer 2013 Magazine 2013; 9. Available at corrdefense.nace.org.

17. Haeseldonckx D., D'haeseleer W. The use of the natural-gas pipeline infrastructure for hydrogen transport in a changing market structure. Int J Hydrogen Energ 2007; 32: 1381–1386.

18. Han Z. Y., Weng W. G. Comparison study on qualitative and quantitative risk assessment methods for urban natural gas pipeline network. J Hazard Mater 2011; 189: 509–518.

19. Hanafy M. Flow-accelerated in gas compression piping. Mater Perform 2015; 54: 46–50.

20. Heidersbach R. Metallurgy and corrosion control in oil and gas production. Wiley, 2011.

21. Heidersbach R. H., Dexter S. C., Griffin R. B., Montemarano J. Marine corrosion. ASM Handbook 1987; 13: 893–926.

22. Hernandez G., Schorr M., Carpio J., Martinez L. Preservation of the infrastructure in the Gulf of Mexico. Corros Rev 1995; 13: 65–79.

23. Hernandez M. A. L., Martinez D., Gonzalez R., Perez Unzueta A., Mercado R. D., Rodriguez J. Corrosive wear failure analysis in a natural gas pipeline. Wear 2007; 263: 567–571.

24. Hummel R. Alternative futures for corrosion and degradation research. Arlington, VA: Potomatic Institute Press, 2014.

25. Inzunza R. G., Valdez B., Schorr M. Corrosion inhibitor patents in industrial applications—a review. Rec Patent Corros Sci 2013; 3: 71–78.

26. Javaherdashti R., Nwaoha C., Tan H. Corrosion and materials in the oil and gas industries. Boca Raton, Florida: CRC Press, 2013.

27. Jo Y. D., Ahn B. J. A method of quantitative risk assessment for transmission pipeline carrying natural gas. J Hazard Mater 2005; 123: 1–12.

28. Kargbo D. M., Wilhelm R. G., Campbell D. J. Natural gas plays in the Marcellus shale: challenges and potential opportunities. Environm Sci Technol 2010; 44: 5679–5684.

29. Lucas E. Energy and technology. The Economist, January 17, 2015.

30. Marcos M., Botana J., Valdez B., Schorr M. Polución y corrosión en ríos contaminados, 3er. Congreso de Ingenieria Civil, Territorio y Medio Ambiente, Zaragoza, 2006 (in Spanish).

31. Melman Y. Gas rig standoff. The Jerusalem Report 2014: 12–15.

32. Mexican Energy Reform. The Economist, November 23, 2013, 38–39.

33. Mokhatab S., William P., James S. Handbook of natural gas transmission and processing. Atlanta, USA: Elsevier, 2006.

34. NACE MR0175/ISO 15156-3. Petroleum and natural gas industries materials for use in H_2S containing environments in oil and gas production. Houston, TX: NACE, 2009.

35. NACE Standard SP 0110. Wet gas corrosion, direct assessment methodology for pipelines. Houston, TX: NACE, 1998.

36. Papavinasam S., Revie R. W. Review of standards for evaluating coatings to control external corrosion of pipelines. Corros Rev 2008; 26: 295–371.

37. Papavinasam S., Attard M., Revie R. W. Evolution of external pipeline coatings for corrosion protection—a review. Corros Rev 2008; 26: 373–438.

38. Park A. Upgrading the disaster. Time Magazine, April 25, 2011. Philippine Corrosion Society. Q&A on corrosion of metals in oil and gas and petrochemical industries. Japan Society of Corrosion Engineering, editor. 2009.

39. Quintero Nuñez M., Valdez B., Schorr M. Effect of H_2S on corrosion in polluted waters. Adv Mater Res 2010; 95: 33–36

40. Quintero Nuñez M., Sanchez-Sanchez C. del C., Garcia-Cueto R., Santillan-Soto N., Ojeda-Benitez S., Velazquez-Limon N. Environmental impact of the Energía Costa Azul LNG terminal at Ensenada, B.C. México. WIT Trans Ecol Environ 2014; 181: 15–24.

41. Raichev R., Veleva L., Valdez B. Corrosión de metales y degradación de materiales. Schorr M, editor. Universidad Autónoma de Baja California, 2009 (in Spanish).

42. Rebak R. B. Sulfidic corrosion in refineries—a review. Corros Rev 2001; 29: 123–133.

43. Roberge P. Corrosion engineering, principles and practice. McGraw Hill, 2009.

44. Sastri V. S. Green corrosion inhibitors: theory and practice, vol. 10. Wiley, 2012.

45. Schorr M., Valdez B. Corrosion of the infrastructure in polluted ports. Corros Eng Sci Technol 2005; 40: 137–142.

46. Schorr M., Valdez B. Pollution and corrosion in the Kishon River. In: The Israeli 7th Conference on Corrosion and Electrochemistry, May 2006.

47. Schorr M., Valdez B. Pollution and corrosion in marine and fluvial shipyards. In: XVI International Materials Research Congress, Symposium XV, Cancun, Mexico, August 2007.

48. Schorr M., Valdez B., Quintero M. Effect of H_2S on corrosion in polluted waters: a review. Corros Eng Sci Technol 2006; 41: 221–227.

49. Sidney T., Dawe R. A. Review of ways to transport NG energy from countries which do not need the gas for domestic use. Energy 2003; 28: 1461–1477.

50. Smith R. Practical natural gas engineering. Pennwell Books, 1990.

51. So A., Valdez B., Schorr M. Materiales y corrosión en la industria de gas natural. In: Valdez B, Schorr M, editors. Corrosión y protección de la infraestructura industrial, capítulo XX. OMNIA Science Monograficos, 2013: 87–102 (in Spanish).

52. Srinivassan S., Eden D. C. Natural gas internal pipeline corrosion. ASM handbook, vol. 13 C. Corrosion: environments and industries. Materials Park, OH: ASM International 2006: 1026–1036.

53. The Natural Gas Authority. The natural gas sector in Israel. Available at: energy.gov.il. Accessed February 2015.

54. Valdez B., Schorr M. Hydrogen sulfide, a dangerous, corrosive agent. Symposium 7, NACE: corrosion and metallurgy. In: XX International Materials Research Congress, Cancun, Mexico, 2011.

55. Valdez B., Schorr M., Quintero M., García R., Rosas N. The effect of climate change on the durability of engineering materials in the hydraulic infrastructure, special issue: relationship of corrosion with climate change. In: Valdez B, Schorr M, guest editors. Corros Eng Sci Technol 2010; 45: 34–41.

56. Valdez B., Schorr M., So A., Eliezer A. LNG regasification plants, materials and corrosion. Mater Perform 2011; 50: 64–68.

57. Valdez B., Schorr M., Zlatev R., Carrillo M., Stoytcheva M., Alvarez L., Eliezer A., Rosas N. Corrosion control in industry. In: Vadez B, Schorr M, editors. Environmental and industrial corrosion. Croatia: INTECH, 2012: 19–53.

58. Veleva L., Castro P., Hernandez-Duque G., Schorr M. The corrosion performance of steel and reinforced concrete in a tropical humid climate. A review. Corros Rev 1998; 13: 235–284.

59. Waard C., Lotz U., Milliams D. E. Predictive model for CO_2 corrosion engineering in wet natural gas pipelines. Corrosion 1991;47: 976–985.

60. Wang X. Advanced natural gas engineering. Largo, Florida: Gulf Publishing, 2009.

61. Wang J., Atrens A. Analysis of service stress corrosion cracking in a natural gas transmission pipeline, active or dormant? Eng Fail Anal 2004; 11: 3–18.

62. Wong-Dickason J., Thomson S. Chemically treating assets in the Bakken formation. Mater Perform 2013; 52: 42–46.

Corrosion Inhibitor Patents in Industrial Applications—a Review

R. G. Inzunza, Benjamín Valdez, and Michael Schorr
Instituto de Ingeniería, Universidad Autónoma de Baja California, CP 21280, Mexicali, Mexico

18.1 Abstract

Corrosion affects the quality of the environment and the durability of the infrastructure assets and industrial equipment. Therefore, it is crucial using corrosion engineering control methods and techniques—in particular safe 'green' (e.g., environmental-friendly corrosion inhibitors that will extend the life of the infrastructure, saving large expenses in materials equipment and structures). This review presents an analysis of patents on corrosion inhibitors developed for aqueous systems, steel-reinforced concrete, acid pickling operations, oil industry, and additives in the formulation of protection coatings.

Keywords: corrosion inhibitor, patents, water systems

18.2 Introduction

Naturally, the corrosion phenomenon appears in numerous technological and industrial applications. Proper material selection is essential to find corrosion-resistant materials (CRMs) suitable to the operating conditions. At the same time, economic and adequate control methods should be employed to extend the useful life of aluminum, copper, steel, and magnesium alloys. Therefore, corrosion inhibitors are used in a wide range of industrial and commercial applications such as cooling water system, reinforced concrete, acid pickling, steam-condensate lines, and handling and storage of electronic and military equipment.

Several corrosion inhibitors have been developed from natural raw materials to protect different metals that are used in quotidian life. Eventually, collaborations of research groups have been investigated about the mechanisms of the corrosion inhibition processes, the optimum operation conditions of corrosion inhibitors, as well as the most appropriate type of molecule for each metal and corrosive media system. The efficiency of a corrosion inhibitor may vary under the same industrial applications because of the diverse environmental conditions presented, the kind of industrial equipment, or the physicochemical properties of the water utilised in these systems.

Corrosion can be controlled by modifying the aqueous environment and by neutralising or removing corrosive agents (e.g., dissolved oxygen [DO]). Corrosion inhibitors decrease the rate of corrosion when added in relatively small quantities to the system under corrosion. Corrosion inhibitors can be classified into the following categories:

- anodic inhibitors, which retard the anodic corrosion reactions by forming passive films
- cathodic inhibitors, which supress cathodic reaction, such as reduction of DO
- mixed inhibitors, which interact with both anodic and cathodic reactions
- adsorption inhibitors, such as amines, oils and waxes, adsorbed on the steel surface, forming a thin protective film, preventing metal dissolution

This article analyses diverse patents and studies of research groups in the world that are dedicated to corrosion control, mainly developing corrosion inhibitors in water systems.

18.3 Corrosion Inhibition in Cooling Water Systems

Evaporation is the main source of cooling in an open recirculation system; as cooling proceeds, evaporation results in an increasing concentration of the dissolved solids in the water requiring blowdown.

There are only three basic designs:

1. Open recirculating systems
2. Once-through systems
3. Closed recirculating systems

Open recirculating systems are the most widely utilised industrial cooling design. These systems consist of pumps, heat exchangers, and a cooling tower. The pumps keep the water recirculating through heat exchangers. It picks up heat and moves it to the cooling tower where the heat is released from the water through evaporation. The water in open recirculating systems undergoes changes in its basic chemistry because of evaporation. The dissolved and suspended solids in the water become more concentrated.

In *once-through systems*, the cooling water passes through heat exchange equipment only once. The mineral content of the cooling water remains practically unchanged as it passes through the system. Because large volumes of cooling water are used, these systems are applied less often than recirculating systems. Seasonal temperature variation of the incoming water can create operational problems. Thermal pollution of lakes and rivers by system discharge is an environmental concern.

In *closed recirculating systems*, the same cooling water is moved in a continuous cycle. First, the water absorbs heat from the process fluids and then releases it in another heat exchanger. An evaporative cooling tower is not included in these systems that are often used for critical cooling applications or when water temperature below ambient is required, as in a chilled water system.

Matulewicz et al. patented a corrosion inhibitor formulation comprising one or more tetrahydrobenzotriazoles and one or more other triazoles. The invention relates to a method for corrosion inhibition of copper and rich copper alloys in the presence of corroding agents by contacting that metal component with a corrosion inhibitor formulated with a blend of tetrahydrotolyltriazole (THTT), benzotriazole (BT), and tolyltriazole (TT). The amount of THTT should be adjusted appropriately, regarding, for example, to the BT/TT rate in terms of how much THTT should be blended with BT/TT to improve the corrosion inhibition. These corrosion inhibitors are added to the cooling water systems, as well as other aqueous and non-aqueous systems.[1]

Grech et al. patented a formulation comprising a Tetrakis (hydroxyorgano) phosphonium salts (THP+) salt and a thio-substituted compound. The formulation is recommended for the treatment of corrosion and metal sulphide scale deposits in aqueous systems. The thio-substituted compound may be a thio-substituted carboxylic acid or salt, a thio-substituted sulphonic acid, a substituted or unsubstituted alkyl or aryl thiol, a thio-substituted heterocyclic compound or a mercaptoethanol. The formulation may further include a surfactant. The surfactant is preferably a cationic surfactant—for example, quaternary ammonium compounds, N-alkylated heterocyclic compounds, or quaternised amido-amines.[2]

Michael A. Silveri patented water treatment chemical formulations containing halogens as a sanitising agent for water treatment. The water treatment chemical formulations contain cyanuric acid and an alkali metal molybdate or silicate as anticorrosion agent.[3] Another invention provides novel amine-based white rust corrosion-inhibiting compounds capable of significantly inhibiting white rust formation in industrial water systems having galvanised metal surfaces. The compounds include those of the formula $[R]_2N-Rlx-N[R]_2$. R includes H_5 mono-hydroxylated alkyl groups or poly-hydroxylated alkyl groups. R1 includes $-[CH_2CH_2-N(RS)-CH_2CH_2]$, oxypropylene, oxyethylene, polyether, the like, or combinations thereof. R2 includes H, alkyl groups, alkylated carboxylates, alkylated sulphonates, mono-hydroxylated alkyl groups, or poly-hydroxylated alkyl groups. X is from 1 to about 20. The invention was patented by Greene et al.[4]

Recently, there are patented methods of corrosion inhibition of copper, nickel, aluminum, zinc, tin, lead, beryllium, carbon steel, various alloys of such metals, and galvanised coatings in evaporative cooling water applications approaching zero liquid discharge that are specifically attacked by cooling water with residuals of corrosive chemistry or ions such as ammonia/ammonium ion, chloride, high TDS, OH^-. The method includes applying azoles inhibitors (such as TTA, BTA, etc.) at residuals of 0.25 mg/L to 200 mg/L or greater (as azoles) to the cooling water application and operating with a combination of high TDS (greater than 2500 mg/L) and high pH (greater than 9.0), while maintaining low total hardness (less than 200 mg/L as $CaCO_3$). The patent was obtained by Duke D.[5]

For closed cooling water systems (e.g., engine coolants), the problems associated with scaling must be taken in account. Heat transfer fluids, which generally composed by water, glycol or glycol-water mixtures, are in contact with one or several metallic parts that are prone to corrosion. A corrosion-inhibiting composition of monocarboxylic acid (octanoic acid), dicarboxylic acid (sebacic acid), triazole, and triethanolamine compound was patented by Belokurova.[6] Burda P. patented a potassium salts inhibitive blend and corrosion protection method for closed cooling water systems. The closed cooling system circulates about thirty-five to forty gallons of corrosion inhibitor per ten thousand gallons of circulating water. The corrosion inhibitor contains potassium molybdate, potassium nitride, and potassium borate in an aqueous solution so that if leakage of the circulated water occurs, the leaked corrosion inhibitor solution does not contaminate the environment. The potassium molybdate and potassium nitrate each constitute about 12 per cent of the aqueous solution. Potassium-based borate at 2 to 2.5 per cent, ethylene glycol preferably at about 10 per cent, and benzotriazole or tolyltriazole of about 0.4 per cent with up to 1 per cent polymeric scale suppressant are also included in the corrosion inhibitor blend.[7] Baker and Christensen patented an all-in-one treatment formulation for water cooling systems that includes a polymeric quaternary ammonium compound poly[oxyalkylene(dialkylimminio)alkylene(dialkylimminio)] salt, which is effective as a continuous biocide/biostat at low levels, an organophosphorous carboxylic acid compound that acts as a scale control agent, dispersant and corrosion inhibitor, as well as molybdate corrosion inhibitor and azole-type corrosion inhibitor. These are provided in high concentrations for storage and shipment and are applied at low level concentration on a continuous basis in a water-cooling tower.[8]

18.4 Corrosion Inhibitors for Steel-Reinforced Concrete

Although steel-reinforced concrete is generally very durable, because of the use of diverse cement formulations, concrete constructions can suffer from visible damage that is attributable to corrosion of the steel reinforcement and sometimes occurs only a few years after the construction has been built, although the life of a steel-reinforced concrete construction is usually planned as 100 years. Reinforced concrete structures under corrosion would require an expensive maintenance program; therefore, the application of effective corrosion inhibitors represents a good alternative to protect the steel reinforcement.

In 1987, the patent of Banks et al. was published with the title 'Inhibiting Corrosion in Reinforced Concrete' in which the vapour phase corrosion inhibitors (VPI) are inserted into reinforced concrete structures, either during their construction, in which case the self-supporting VPI insert, which is independent of the metal reinforcement, may be in the form of a spacer for spacing the reinforcing rods; or after their formation finished by insertion of the fluid or solid VPI in holes bored in the pre-existing structure. Preferred VPI materials are cyclohexylamine nitrites, benzoates, and carbamates.[9] In 1997, Chandler et al. patented a corrosion inhibitor formulation for use in reinforced concrete structures; the inhibitor reduced the rate of corrosion in metallic reinforcing rods placed within the structures. The formulation comprises a mixture of benzoic acid, aldonic acid, and a triazole such as benzotriazole or tolyltriazole.[10] Malric and Lutz patented a potassium monofluorophosphate as a corrosion inhibitor. The potassium monofluorophosphate in aqueous solution is contacted with a surface of the reinforced concrete to facilitate penetration of the potassium monofluorophosphate into the concrete.[11] A composition for protecting steel-reinforced concrete against corrosion of the steel reinforcement was patented by Standke and McGettigan in 2004.[12] The corrosion inhibitor contains a component A, at least one organosilane or organosiloxane of the formula (I)

$$R\text{-}SiR^1_x(O)_yR^2_z, \qquad (I)$$

where R is a linear or branched alkyl group having from three to twenty carbon atoms; R^1 is a linear or branched alkyl group having from one to four carbon atoms; R^2 is a linear or branched alkoxy group having from one to four carbon atoms or a hydroxy group, where groups may be identical or different; x is 0, 1, or 2; y is from 0.0 to 1.5; z is 0, 1, 2 or 3; and $(x + 2y + z) = 3$, wherein the composition is an oil-in-water emulsion; and an oil phase of the oil-in-water emulsion comprises at least one of the component A (corrosion inhibitor), a component B, a component C, and a component D, where the component B comprises at least one aminosilane or aminosiloxane of the formula (II)

$$R^4_2N\text{-}(R^5\text{-}NR)_a\text{-}R^6\text{-}Si(R^7)_bO_c(OR^8)_d, \qquad (II)$$

where R is a hydrogen atom or a linear or branched alkyl group having from one to twenty carbon atoms, and groups R^4 are identical or different; R^5 is a linear or branched alkylene group having from one to twenty carbon atoms; R^6 is a linear or branched alkylene group having from one to twenty carbon atoms; R^7 is a linear or branched alkyl group having from one to four carbon atoms; R^8 is a hydrogen atom or a linear or branched alkyl group having from one to four carbon atoms, and groups R^8 are identical or different; a is 0, 1, 2, 3, or 4; b is 0, 1, or 2; c is from 0.0 to 1.5; d is 0, 1, 2, or 3; and $(b + 2c + d) = 3$; the component C comprises at least one carboxylic acid or salt of a carboxylic acid; and the component D comprises at least one aminoalcohol $(HO\text{-}CH_2\text{-}CH_2\text{-}NR^9_2)$, where R^9 is a hydrogen atom or a linear or branched alkyl group having from one to twenty carbon atoms. Groups R^9 are identical or different.

A 50:50 combination sodium benzoate and alkali metal sebacate was used as a migrating inhibitor (VCI) in the polymer fibre in post-tensioning cables.[13]

18.5 Corrosion Inhibitors for Acid Solutions

In the construction of oil, gas, or water wells, pickling acid washing, matrix acidising, and acid fracturing are operations performed during the production stimulation processes. The shale-gas industry requires the use of corrosion inhibitors to reduce the rate of acid attack on metal components and to protect wellbore tubular goods. Drilling and construction of a well is an expensive activity; special alloys for pipelines and tools, labour costs, energy, machinery, and chemicals are required along the operation. The cost of corrosion inhibitors is high and represents a significant portion of the total costs.

Recently, Malwitz et al. patented their invention relating with compositions and methods for environmentally friendly oil and gas industry corrosion inhibitors. More specifically, the invention relates to compositions including a quaternary nitrogen-containing corrosion inhibitor (figure 1) that meets current 'green' chemistry regulations that have reduced environmental impact. The invention has particular relevance to corrosion inhibitors that fulfil three of the North Sea criteria benchmarks of biodegradation, bioaccumulation, and toxicity.

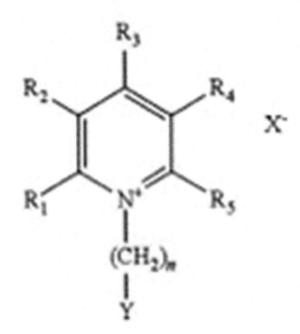

Figure 1. General formula of a quaternary nitrogen-containing corrosion inhibitor.

The variables in the general formula represent the following: (I) n is an integer from 1 to 4; (II) R_1, R_2, R_3, R_4, and R_5 are each independently selected from H, $-(CO_2H)_w$ (i.e., di-carboxylic acid), $-(CO_2R_6)X$ (i.e., di-ester), $-(C(=O)NR_7R_8)_w$ (i.e., di-amido), $-C(O)NR_7R_8$, $- N(H)C(=O)R_8$, tetrazolyl, substituted tetrazolyl, alkoxy, dialkoxy, alkyl, substituted alkyl, dialkyl, substituted dialkyl, amine, substituted amine, and combinations thereof; wherein R_6 is a C_1-C_8 alkyl or phenyl, R_7 is H or a C_1-C_4 alkyl, R_8 is H or a C_1-C_4 alkyl, W is 1 or 2, and X is 1 or 2; (III) Y is selected from napthyl, benzyl, anthracyl, phenanthrinyl, substituted napthyl, substituted benzyl, substituted anthracyl, substituted phenanthrinyl, and combinations thereof; and (IV) X^- is a counter ion with a charge sufficient to balance the positive charge on the parent compound of the general formula.

The corrosion inhibitor formulation is introduced into the well treating acid at a concentration sufficient to coat the well tubular and equipment. The dosage rate of the acid corrosion inhibitor formulation is dependent on the metallurgy of the well being protected, type of acid used, acid concentration, duration of acid exposure in the well, well temperature, well pressure, and the presence of other chemistries or additives (i.e., external intensifiers, iron control agents, anti-sludge materials, etc.). This invention is claimed to be effective in inhibiting corrosion in several metals and alloys and in a variety of well treating acids used in the oilfield. Types of metallurgy include, for example, mild steels (such as N80, L80, J55), high chrome steels (e.g., 13Cr85), and coil tubing (e.g., CT900). Representative well treating acids include hydrochloric acid (HCl), hydrofluoric acid (HF), mixtures of HCl and HF (i.e., mud acid), acetic acid, formic acid and other organic acids, and anhydrides.[14]

A corrosion inhibitor was patented by Singh et al., and its composition essentially comprises alkarylated polyalkyl pyridinium salts of the general formula (figure 2) (where R,R'.R" is a linear or branched C1 - C5 alkyl chain, R''' is benzyl group, X is an anion preferably halogen) containing at least one or more than one alkanol, a surfactant, an alkynol; such composition inhibits the over pickling of the ferrous metal and its alloys during the hydrochloric acid/sulphuric acid pickling.[15]

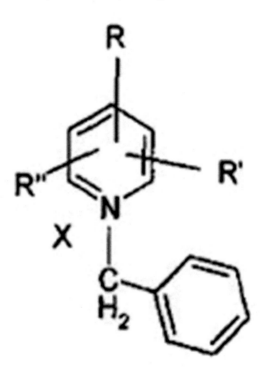

Figure 2. General formula of alkarylated polyalkyl pyridinium salts.

Schacht et al. patented their invention related to sulphuric/nitric blended acid cleaners, which employ the use of ethoxylated amines and/or ethoxylated alcohols as corrosion and stain inhibitor in the vapour phase for cleaning metal and other surfaces, particularly stainless steel.[16]

McCormick et al. patented an acid inhibitor formulation for metal cleaning and/or pickling. This acid inhibitor concentrate contains water, at least one polyamino-aldehyde resin such as a quaternised polyethylenepolyamine-glyoxal resin, and at least one compound selected from the group consisting of acetylenic alcohols, ethoxylated fatty amines, ethoxylated fatty amine salts, and aldehyde-releasing compounds (such as hexamethylenetetramine). The resulting concentrates form useful metal cleaning and pickling solutions when combined with aqueous acid, wherein such solutions, when contacted with a metal surface, are effective in removing scale, smut, and other deposits from the metal surface but exhibit a reduced tendency for the aqueous acid to attack or etch the metal itself.[17]

Cassidy et al. invented and patented a method for corrosion inhibition for subterranean applications, and more particularly, to corrosion inhibitor compositions comprising products of a reaction between aldehydes and amides and methods related thereto.[18] The corrosive environment preferably comprises an acidic environment having a pH of 5 or below. In another aspect of the present invention, at least one amide involves a primary amide or a secondary amide. Preferably, at least one amide that has the general structure showed in figure 3,

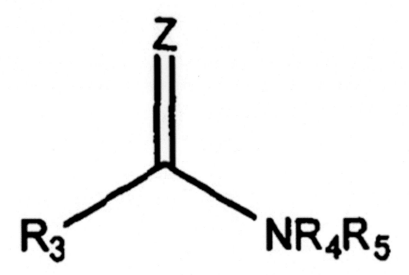

Figure 3. General formula for amide.

where in R_3 is alkyl, trihaloalkyl, alkenyl, alkynyl, aryl, aralkyl, cycloalkyl, heterocyclyl, heteroaryl, or heteroaralkyl; and wherein R_4 and R_5 are independently hydrogen, alkyl, alkenyl, alkynyl, aryl, aralkyl, cycloalkyl, eterocyclyl, heteroaryl, or heteroaralkyl; and wherein at least one of R_4 or R_5 is H. The R_4 and R_5 may be H and Z is O.

The aldehyde general structure formulas are shown in figure 4,

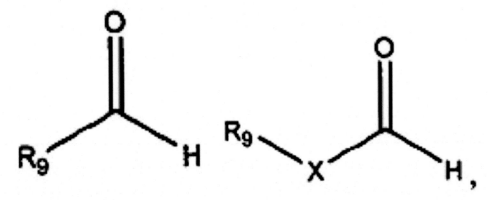

Figure 4. General structural formulas for aldehyde molecules.

where in R_9 is alkyl, alkenyl, alkynyl, aryl, aralkyl, cycloalkyl, heterocyclyl, heteroaryl, or heteroaralkyl; and wherein X is $(CH!)$!, $CH = CH$, or $C \equiv C$; wherein n is an integer ranging from 1 to about 10.

Gino et al. patented an aqueous organic acid formulation containing a terpene as corrosion inhibitor intensifier that is especially suitable for use in acidising subterranean formations and wellbores. The composition substantially reduces the corrosive effects of the acidic solution in contact with metallic surfaces. Appropriate terpenes for the formulation include carotene, limonene, pinene, farnesene, camphor, cymene, and menthol.[19] Vorderbruggen et al. patented an acid corrosion inhibitor. The inhibitor combines cinnamaldehyde and an organo-sulphur compound. The inhibitor provides a reduced rate of corrosion and fewer instances of pitting than inhibitors that include cinnamaldehyde alone.[20] Choudhary et al. patented their invention title as 'Methods of Inhibiting Corrosion of a Metal Surface in a Well or Pipeline Are Provided'. The methods include the steps of (a) forming a fluid of an aqueous acid solution and inulin and (b) introducing the fluid into the well or pipeline. The methods have wide application in various kinds of operations involved in the production or transportation of oil and gas, such as acid stimulation or remedial treatment in a pipeline. The inulin acts as corrosion inhibitor. Preferably, the inulin or source of inulin (extracted from the roots of the chicory plant) is in the form of a particulate prior to the step of combining with the aqueous acid solution. Most preferably, the particle size of the corrosion inhibitor is in the range of a powder. The particulate may be suspended in a liquid for ease handling and mixing prior to the step of combining. For example, the corrosion inhibitor can be suspended in an oil phase. The inulin is combined with the aqueous acid solution in an amount in the range of from about 0.01 per cent wt/vol to about 5 per cent wt/vol of the aqueous acid solution. More preferably, the inulin is combined with the aqueous acid solution in an effective amount to provide at least measurable corrosion inhibition for the metal to be contacted by the well fluid.[21]

18.6 Corrosion Inhibitors for the Oil Industry

The oil industry involves four main processes: (I) primary production, (II) secondary production, (III) refining, and (IV) storage. Corrosion problems arise in the primary production because of the presence of water that accompanies the oil. The water can contain various corrosive agents such as carbon dioxide, hydrogen sulphide, organic acids, chlorides, and sulphates. The wells containing hydrogen sulphide are known as sour wells, and the wells devoid of hydrogen sulphide are known as sweet wells. Sour wells are very corrosive. When the oil-water ratio is suitable, the crude oil by itself can be protective against corrosion.

Most inhibitors consist of organic compounds containing nitrogen such as aliphatic fatty acid derivatives, imidazolines, quaternary ammonium compounds, complex amine mixtures based on abietic acid, petroleum sulphuric acid salts of long-chain diamines, other salts of diamines, and fatty amides of aliphatic diamines.[22]

Polley patented an anti-corrosion oil formulated with a mineral lubricating oil, cyclohexylamine salt of a C8-C24 fatty acid (preferably oleic acid) and ricinoleic acid, the amount of ricinoleic acid being 0.1 to 10 parts by weight based on the cyclohexylamine salt, and preferably 0.5 to 3 parts by weight. In examples, a mineral oil of viscosity 55 S.U.S/210 F and specific gravity (60/60) 0.939 contains 1 per cent ricinoleic acid and 0.5 to 3 per cent of cyclohexylamine oleate.[23]

Al-Zahrani patented a method of mixing a corrosion inhibitor in an acid-in-oil emulsion. This invention provides a method to enhance hydrocarbon recovery when acidising a wellbore in a carbonate formation and to inhibit the corrosion of oil well tubing by using an acid-in-oil emulsion. An acid is added initially to an emulsified oil so that only the acid is contained within the oil, or a coal tar distillate, such as naphtha, gasoline, kerosene, or carbon tetrachloride, and suitably mixed, will form an emulsion. The corrosion inhibitor is then added to the acid-in-oil emulsion and mixed with it so that it forms the external phase, with the acid in the emulsified oil forming the internal phase. The disposition of the corrosion inhibitor in the external phase of the emulsion enhances its ability to contact and dispense on the metal surfaces of the oil well tubing in a much more efficient manner as the emulsion is delivered by pumping downwardly into the formation. Thus, when a droplet of the acid-in-oil emulsion of the present invention is viewed in cross-section, the innermost phase is the acid that is encapsulated by the oil, and the corrosion inhibitor forms the external phase. Accordingly, the corrosion inhibitor of the present invention must be immiscible with the acid but, instead, will mix with the entire emulsion from the outside to form an external phase. The corrosion inhibitor functions by adsorbing onto the surfaces of the coil tubing and the well tubing providing a protective barrier between it and the acid. The presence of the corrosion inhibitor in the external phase of the emulsion expedites the adsorption of the corrosion inhibitor onto the steel surfaces to protect the well tubing.[24] Leinweber et al. patented low-toxic, water-soluble, and biodegradable corrosion inhibitors of metal salts of $CH_3SCH_2CH_2CH(NHCOR)COOH$ containing anionic and cationic surfactants.[25, 26] Beloglazov et al. disclosed an antipyrine derivative as an inhibitor against microbiological (myxromycete) corrosion of equipment made of carbon steel and alloy steel with cadmium coating.[27] Meyer and Monk patented a corrosion inhibitor formulation in oil and gas applications that comprises a mixture of substituted amines and fatty acids and branched dodecylbenzenesulphonic acid (DDBSA). The mixture is derived from a reaction between a complex mixture of alkanolamines with various fatty acids. In an aspect, the general reaction neutralises the raw materials, forming a complex mixture of salts of the alkanolamines with the fatty acids. At least one fatty acid contains at least one selected from trimeric C18 unsaturated fatty acid; dimer acids, polymerised tall-oil fatty acids; at least one component of a crude tall oil composition; and any combination of the foregoing. At least one alkanolamine comprises at least one N, N-dimethylethanolamine. At least one substituted alkylamine comprises dimethylethanolamine. The corrosion inhibitor also contains a solvent that could be methanol, ethanol, propanol, isopropanol, butanol, isobutanol, aromatic hydrocarbons, monoethyleneglycol, ethyleneglycolmonobutylether, and combinations thereof. Furthermore, the corrosion inhibitor contains a dispersant; at least one dispersant comprises at least one oxyalkylate polymer.[28]

18.7 Vapour Phase Corrosion Inhibitors (VCIs)

Vapour phase corrosion inhibitors are best known to protect metallic articles and equipment during storage and transportation. There are many investigations on corrosion inhibition studies by aliphatic amines, salicylic amines, and their salts as vapour phase corrosion inhibitors for various industrial metals and alloys.[29-30] Benzotriazole has been reported as an effective corrosion inhibitor for copper and its alloys.[31] A vapour phase corrosion inhibitor-desiccant composite consists in silica gel granules coated with a vapour phase corrosion inhibitor component. The corrosion inhibitor component is selected from a formulation comprising anhydrous molybdates such as ammonium dimolybdate, sodium molybdate, and amine molybdates mixed with benzotriazole and sodium nitrate, or from a formulation comprising amine benzoates, amine nitrates, and benzotriazole. The composites can be impregnated into foam, extruded with polyolefin films, which can additionally be laminated with a metalised second film or encapsulated in an air-permeable container. The corrosion inhibitor formulations have vapour pressures, which provide ongoing corrosion protection for susceptible articles situated favourably with respect to the composite. The corrosion inhibitor was patented by Foley et al.[32] Bradley et al. patented a biodegradable resin product consisting essentially of a polymeric resin of polyethylene, starch, polyesters such as polylactic acid, or other suitable polyesters. In admixture with the resin is a particulate vapour phase corrosion inhibitor selected from amine salts, ammonium benzoate, triazole derivatives, tall oil imidazolines, alkali metal molybdates, alkali dibasic acid salts, and mixtures thereof, and is present in an amount ranging from between about 1 per cent to 3 per cent by weight of the polymeric resin.[33] An effective corrosion inhibitor polymer composition including an interceptor for an acid gas combined with known VCI ingredients produces a more effective VCI than the prior art composition; this was patented by Kubic et al. The composition in macro granular form may be held in a receptacle that is sealed in a container in which ferrous articles are protected. When the composition, greatly diluted, is injection- or blow-moulded or extruded as film that contains more than 100 ppm of a 2,4-6-tri-substituted phenol and uniformly dispersed small amounts of solid micron-sized particles of interceptor, an alkali metal nitrite and a solid adjuvant to aid in dispersing the solid particles, the film is transparent and has a smooth surface. All solid particles in the material have a primary particle size smaller than 53 μm, which makes it possible to obtain the uniform dispersion, transparency, and smooth surfaces of an article such as film in which a ferrous metal object is wrapped prior to shipment; or a capsule that is filled with the VCIs and placed with ferrous metal objects in a sealed container.[34] Lozano et al. patented a vapour phase corrosion inhibitor; the composition of the present application is particularly suitable for incorporation within plastic wraps in the form of coated woven products, laminated films, and blown films. Preferably, the final composition of the plastic wrap comprises between 0.13 per cent and 1.25 per cent by weight alkali metal nitrite, between 1 per cent and 5.63 per cent by weight alkali metal benzoate, and between 0.25 per cent and 3.13 per cent by weight alkali metal molybdate. Instead of one or more of the alkali metal compounds, one or more of the corresponding alkaline earth metal compounds may be used. In a more preferred embodiment, the composition for a 1 mil (i.e., 10^{-03} inch), includes 0.6 per cent sodium nitrite, 1 per cent sodium molybdate, and 4.5 per cent sodium benzoate.[35]

Reinhard et al. patented the invention related to new substance combinations as vapour phase corrosion inhibitors for protecting a broad range of customary utility metals, including iron, chromium, nickel, tin, zinc, aluminium, copper, magnesium, and their alloys against corrosion in humid climates. The substance combinations comprise (I) at least one C6 to C10 aliphatic monocarboxylic acid, (II) at least one C6 to C10 aliphatic dicarboxylic acid, and (III) a primary aromatic amide. Preferably, they further comprise (IV) an aliphatic ester of hydroxybenzoic acid—in particular, of 4-hydroxybenzoic acid and/or (V) a benzimidazole, in particular a benzimidazole substituted on the benzene ring.[36] Two years later, also Reinhard et al. patented an invention related to substance combinations prepared with (I) at least one substituted, preferably polysubstituted, pyrimidine; (II) one or more monoalkylurea; (III) at least one C3 to C5 aminoalkyldiol; and, optionally, (IV) a benzotriazole, preferably a benzotriazole, which is substituted on the benzene ring. The components may be mixed together or dispersed in water or pre-mixed in a solvent that is miscible in any ratio with mineral oils and synthetic oils, such as, for example, a phenyl alkyl alcohol or an alkylated phenol. Such substance combinations can be used as vapour phase corrosion inhibitors in packagings or during storage in closed spaces for protecting customary utility metals, such as iron, chromium, nickel, tin, zinc, aluminium, copper, and alloys thereof against atmospheric corrosion.[37]

18.8 Corrosion Inhibition in Coatings

Hobbins et al. patented a corrosion inhibitor of copper-containing metals. The treatment of copper-containing bodies with 5-methyl benzimidazole yields excellent protection from oxidation and/or corrosion of the treated copper-containing body. Additionally, this treatment does not interfere with a subsequent soldering of the copper surfaces.[38]

Gaglani patented corrosion-inhibiting compositions and methods for applying these compositions to metal surfaces to inhibit rusting and formation of blisters wherein the compositions comprise at least one aminocarboxylate salt (figure 5),

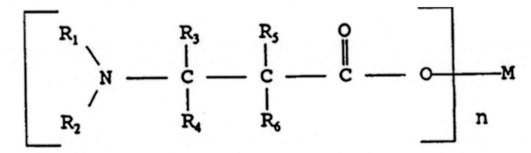

Figure 5. Aminocarboxylate salt.

wherein M is a metal ion and $n = 2$–4, preferably an alkali earth or transition metal, and most preferably Zn, Sn, or Ca; R_1 and R_2, independent of each other, is H, C_1-C_{20} alkyl, alkylene, and where R_1 and R_2 may also combine to form a fused cycloalkyl group, cycloalkenyl group, or a heterocyclic group containing O, N, or S as part of the ring; R_3, R_4, R_5 and R_6, independently of each other, are hydrogen, lower alkyl or lower substituted alkyl, phenyl, substituted phenyl, cycloalkyl having five to six carbon atoms, benzyl, or substituted benzyl.

- A pigment: The pigment can be iron oxide, titanium dioxide, magnesium silicate, clays, zinc oxide, calcium silicate, calcium carbonate and the like.
- A binder. For water-based compositions such as latex paints, the binder is a latex polymer-water emulsion or dispersion in which the polymer can be an acrylic, vinyl acrylic, or polyvinyl alcohol polymer. The solvent is water soluble and can be principally water itself. For oil-based compositions or paints employing fatty acids or fatty oils, polybasic acids, and polyhydric alcohols, such as alkyd paints, the binder can be an alkyd resin containing 50 to 70 per cent solids. An appropriate solvent can be selected from the group consisting of mineral spirits, naphtha, xylene, toluene, butanol, and 2-butoxyethanol, and the like. The oil-based compositions can further comprise additives such as a metal drier and an anti-skinning agent.
- The solvent.[39]

Hughes et al. patented a corrosion inhibitor composition for use in coating products applied on metal surfaces, or in metal containers, to provide corrosion resistance to such metals. The corrosion inhibitor comprises at least one alkoxylated phosphate ester and one or more borate ester. The corrosion inhibitor composition optionally contains inorganic metal salts. There is also provided an end-use coating composition with the content of the components mentioned above and a method of providing corrosion resistance to a metal surface by applying the coating composition thereto.[40] Besing et al. patented a non-carcinogenic corrosion-inhibiting additive based on the use of anodic and cathodic corrosion inhibitors, in combination with a metal complexing agent that increases the solubility of the corrosion inhibitor additive. The resulting product of the present invention provides protection against both localised pitting corrosion and general corrosion. The metal complexing agent is preferably a water-soluble organic acid salt and/or a water-soluble inorganic acid salt. Particularly useful metal complexing agents are selected from the group consisting of citrate, gluconate, polyphosphate, tartrate, β-diketonates, α -hydroxy acids, D-fructose, L-sorbose, and mixtures. The metal complexing agent should be preferably from 0.3 to 0.7 with respect to the mole fraction of the combined anodic and cathodic corrosion inhibitors. The additive may be applied in any manner known in the art including as a conversion coating, or applied as a primer, adhesive, epoxy, paint, organic sealant, sealer for anodised aluminium, additive for recirculating water system, or the like. The use of the corrosion inhibitor additive includes aerospace, automotive, architectural, packaging, electronics, and marine applications.[41]

Von patented an invention that relates to coatings, such as paints containing tobacco products and the use thereof as corrosion inhibitors. The tobacco products include various forms of tobacco such as dried tobacco leaves, stems, dust, liquid extracts, etc. that can be added to the coatings. The invention further relates to treatment methods and compositions for surface treatments such as descaling, pickling, and removing surface deposits and corrosion products.[42] Monzyk et al. patented metal ferrate compounds as inhibiting additives in primer coatings, employing a novel method that increased ferrate compounds solubility in water.[43] Arsenault et al. patented their invention related to corrosion inhibitors and, more particularly, to a protective coating comprising a non-tungstate anodic corrosion inhibitor consisting of zinc molybdate and a cathodic corrosion inhibitor consisting of cerium citrate, which is effective for use on aluminium alloys having relatively high amounts of zinc.[44] Matzdorf et al. patented their invention related to galvanic aluminium alloy powder pigments coated with a semi-conducting corrosion-inhibiting oxide and the process for preparing coated powder-pigments in combination with film-forming binders for coating metal substrates to inhibit corrosion. The coated aluminium alloy powder pigments are electrically active aid, preventing corrosion of metals that are more cathodic (electropositive) than the aluminium alloy pigments.[45]

18.9 Corrosion Inhibitors in Other Applications

A cleaning composition based on an aqueous or non-petroleum solvent, and useful for cleaning exterior surfaces such as aircraft exterior surfaces and other metal, glass, rubber, and polymer surfaces possesses solvent-like properties with respect to greasy soils; inhibits corrosion and degradation of rubber; is bio-degradable; forms a stable emulsion with water; remains optically clear and stable at multiple dilutions with water; and conforms to MIL-PRF 87937D. The composition, which includes fatty acid methyl esters, ethoxylated alcohols having an HLB ranging from about 10 to about 14, at least one alkyl polyglycoside having an HLB ranging from about 10 to about 14, alkali metal silicate, from 0.01 to 1 part by weight of an alkali metal hydroxide, from 0.1 to 1 part by weight of a phosphonate-functional alkyl sodium siliconate; from 0.1 to 1 part by weight of an aliphatic phosphate ester; and from 0.1 to 2 parts by weight of a modified carboxylic acid derivatives corrosion inhibitor in an amount effective to prevent corrosion on aluminium, magnesium, titanium, and steel; and water, was patented by Britton.[46] Varughese patented a packaged electronic device including a conductive structure and an encapsulant. The encapsulant has chlorides and a negatively charged corrosion inhibitor for preventing corrosion of the conductive structure. The negatively charged corrosion inhibitor comprises hippurate ions, acetates, formates, tartarates, phosphates, silicates, and nitrites. The negatively charged corrosion inhibitor is formulated with at least 150 parts per million of the encapsulant and has a concentration of five or more times that of a concentration of the chlorides in the encapsulate,[47] a fast-curing polymeric resin, a conductive filler, and one or more near-infrared absorbing additives and, optionally, an oxygen scavenger or corrosion inhibitor or both, and other additives such as reactive or non-reactive diluents, inert fillers, and adhesion promoters. The final product can be conductive, resistive, or anisotropically conductive. In another embodiment, this invention is a method for improving the cure speed of a formulation by exposing it to a near-infrared energy source patented.[48]

A patented printed circuit board with long-term reliability consists of corrosion-inhibiting layers containing benzotriazole or nitrophenylhydrazine.[49] Chang et al. patented sarcosine and its salts as corrosion inhibitors during cleaning after chemical mechanical polishing.[50]

18.10 Current and Future Developments

The application of corrosion inhibitors under particular operating conditions of industrial plants and equipment depends largely on its chemical composition and stability, its corrosion-prevention efficiency, and economic considerations. The importance and relevance of the corrosion inhibitors technology are evident by numerous patents gathered, illustrated and evaluated during the analysis performed in this review.

18.11 References

[1] W. N. Matulewicz, P. Vogt and J. Milawski, 'Corrosion inhibitor compositions comprising tetrahydrobenzotriazoles and other triazoles and methods for using same', U.S. Patent 20120308432 A1, 2012.

[2] J. Grech and C. R. Jones, 'Formulation for corrosion and scale inhibition', WO Patent 2004083131 A1, 2004.

[3] M. Silveri, 'Corrosion inhibitor', U.S. Patent 6811747 B2, 2004.

[4] N. Greene and S. Kidambi, 'Functionalized amine-based corrosion inhibitors for galvanized metal surfaces and method of using same', EP Patent 2099952 A1, 2009.

[5] D. Duke and J. Kubis, 'Cooling water corrosion inhibition method', U.S. Patent 7955553 B2, 2011.

[6] I. N. Belokurova, 'Corrosion inhibiting antifreeze composition for anticorrosion protection of metals surfaces in cooling systems', RU Patent 2362792, 2009.

[7] P. Burda, 'Corrosion inhibition of closed cooling water auxiliary system for nuclear power plants', U.S. Patent 4968478 A, 1990.

[8] G. Baker and R. Christensen, 'Composition useful as corrosion inhibitor, anti-scalant and continuous biocide for water cooling towers and method of use', U.S. Patent 4719083 A, 1988.

[9] L. Banks and H. Hosgood, 'Inhibiting corrosion in reinforced concrete', WO Patent 1987/006958, 1987.

[10] C. Chandler, A. Furman, L. Gelner, M. Kharshan, B. Miksic, and B. Rudman, 'Corrosion inhibitor for reducing corrosion in metallic concrete reinforcements', U.S. Patent 5597514 A, 1997.

[11] B. Malric and T. Lutz, 'Potassium monofluorophosphate as a corrosion inhibitor', U.S. Patent 6596197 B2, 2003.

[12] B. Standke and E. McGettigan, 'Corrosion inhibitor for steel-reinforced concrete', U.S. Patent 6685766 B2, 2004.

[13] B. Miksic, A. Furman, M. Kharshan, and J. Jackson, 'Composition and method for preserving post-tensioning cables in metal reinforced concrete structures', U.S. Patent 7541089, 2009.

[14] M. Malwitz and D. Woloch, 'Environmentally friendly corrosion inhibitor', U.S. Patent 20130112106 A1, 2013.

[15] R. S. Singh, S. Kumar, and A. Ashutosh, 'Corrosion inhibiting compositions for hydrochloric acid / sulfuric acid pickling baths', IN Patent 00819, 2008

[16] P. F. Schacht and E. V. Schmidt, 'Aqueous acid cleaning, corrosion and stain inhibiting compositions in the vapor phase comprising a blend of nitric and sulfuric acid', WO Patent 2012093372 A3, 2013.

[17] D. McCormick and T. Smith II, 'Acid inhibitor compositions for metal cleaning and/or pickling', U.S. Patent 20090032057 A1, 2009.

[18] J. M. Cassidy and C. Kiser, 'Corrosion inhibitor compositions comprising reaction products of aldehydes and amides and related methods', WO Patent 2012072986 A1, 2012.

[19] A. Gino, A. P. Dilullo, and R. James, 'Corrosion inhibitor intensifier and method of using the same', EP Patent 1724375 A2, 2006.

[20] M. Vorderbruggen and D. Williams, 'Acid corrosion inhibitor', U.S. Patent 6117364 A, 2000.

[21] Y. K. Choudhary, A. Sabhapondit, and D. Ranganathan, 'Inulin as corrosion inhibitor', U.S. Patent 20120238479 A1, 2012.

[22] V. S. Sastri, 'Corrosion Inhibition: Theory and Practice', John Wiley and Sons, 2011.

[23] P. T. Stanley, 'Improved oil compositions', GB Patent 824405 1959.

[24] A. A. Al-Zahrani, 'Method of mixing a corrosion inhibitor in an acid-in-oil emulsion', U.S. Patent 8039422 B1, 2011.

[25] D. Leinweber and M. Feustel, 'Corrosion inhibitors containing anionic surfactants', EP Patent 2031094, 2009.

[26] D. Leinweber and M. Feustel, 'Corrosion inhibitors containing cationic surfactants', EP Patent 2031095, 2005.

[27] J. M. Wilson, J. M. Cassidy, and C. E. Kiser, 'Corrosion inhibitor intensifier compositions and associated methods', U.S. Patent 7960316 B2, 2011.

[28] G. Meyer and K. Monk, 'Corrosion inhibitors for oil and gas applications', U.S. Patent 20120149608, 2012.

[29] D. Zang, L. X. Gao, and G. D. Zhou, 'Polyamine compounds as a volatile corrosion inhibitor for atmospheric corrosion in mild steel', Mater. Corrosion, 2008, 58, 594–598.

[30] U. Rammelt, S. Koehler, and G. Reinhard, 'Use of vapor phase corrosion inhibitors in packages for protecting mild steel against corrosion', Corrosion Sci., 2009, 51, 921–925.

[31] N. Bellakhal and M. Dachraoui, 'Study of the benzotriazole efficiency as corrosion inhibitor for copper in humid air plasma', Mater. Chem. Phys., 2004, 85, 366–369.

[32] J. Foley, B. Miksic, and T. Tzou, 'Vapor phase corrosion inhibitor-desiccant material', U.S. Patent 5344589 A, 1994.

[33] S. J. Bradley, C. Chandler, and B. Miksic, 'Biodegradable vapor corrosion inhibitor products', U.S. Patent 6028160 A, 2000.

[34] D. A. Kubik, V. Boris, E. Y. Lyublinski, and B. Nygaard, 'Corrosion inhibiting composition and article containing it', U.S. Patent 7270775 B2, 2007.

[35] E. Lozano and J. A. Shipley, 'Vapor phase corrosion inhibitors', U.S. Patent 6033599 A, 2000.

[36] G. Hahn, G. Reinhard, and P. Neitzel, 'Vapor phase corrosion inhibitors and method for their production', U.S. Patent 7824482 B2, 2010.

[37] G. Hahn, P. Neitzel, and G. Reinhard, 'Compositions of vapour phase corrosion inhibitors, method for the production thereof and use thereof for temporary protection against corrosion', U.S. Patent 20110198540 A1, 2011.

[38] N. Hobbins and R. F. Roberts, 'Copper-containing articles with a corrosion inhibitor coating and methods of producing the coating', EP Patent 0085701 B1, 1982.

[39] K. D. Gaglani, 'Aminocarboxylate salts as corrosion inhibitors in coating applications', WO Patent 1991004952 A1, 1991.

[40] J. Hughes and H. McNamee, 'Corrosion inhibitors', WO Patent 2013116191 A1, 2013.

[41] A. Besing, P. Bhatia, X. Chang, T. Garosshen, M. Jaworowski, M. Kryzman, F. Lamm, X. Tang, X. Yu, and W. Zhang, 'Corrosion inhibiting additive and corrosion inhibiting coating', EP Patent 1493846 A1, 2005.

[42] F. J. Von, 'Coatings including tobacco products as corrosion inhibitors', WO Patent 2008151028 A3, 2009.

[43] B. F. Monzyk, J. A. Ford, J. T. Stropki, D. N. Clark, V. V. Gadkari, and K. P. Mitchel, 'Corrosion resistant primer coating comprising corrosion inhibiting additives and nonaqueous resins', WO Patent 2010045657, 2010.

[44] S. Arsenault, J. Beals, and M. Jaworowski, 'Corrosion inhibiting coating composition for a Zn-containing aluminum alloy', EP Patent 2011899 B1, 2012.

[45] C. Matzdorf and W. Nickerson, 'Aluminum alloy coated pigments and corrosion-resistant coatings', WO Patent 2013015880 A1 2013.

[46] C. Britton, 'Aircraft cleaner formula', U.S. Patent 20060166854 A1, 2006.

[47] V. Mathew, 'Encapsulant with corrosion inhibitor', U.S. Patent 20130193576 A1, 2013.

[48] D. McCormick, J. Nowicki, C. Cheng, and W. O'Hara, 'Compositions for use in electronics devices', U.S. Patent 20080054227 A1, 2008.

[49] H. Sato, 'Printed circuit boards with excellent solder wettability and long-term reliability in electrical and mechanical connection to electronic components', JP Patent 109076, 2008.

[50] S. Y. Chang, 'Sarcosine compound used as corrosion inhibitor', U.S. Patent 184287, 2009.

CHAPTER 19

Corrosion Mitigation in Water Systems

R. García,[1] B. Valdez,[1] M. Schorr,[1] and A. Eliezer[2]

[1]Laboratorio de Materiales, Minerales y Corrosión, Instituto de Ingeniería, Universidad Autónoma de Baja California, CP 21280, Mexicali, México.
[2]Sami Shamoon College of Engineering, Beer Sheba, Israel

19.1 Abstract

Environmental quality, worldwide water scarcity, and clean energy have been established today as central disciplines in modern science, engineering, and technology. They are already being linked to the critical problems of climate change, global warming, and greenhouse gas emissions, all interrelated phenomena.[1-2] Furthermore, it is now generally accepted that corrosion and pollution are harmful processes that are interrelated, since many pollutants accelerate corrosion, and corrosion products such as rust also pollute bodies of water. Both are pernicious processes that impair the quality of the environment, the efficiency of industry, and the durability of the water infrastructure assets. In this time of energy crisis and economic turmoil, it is essential to save water for a thirsty worldwide population of seven billion. Several critical sectors of the water supply infrastructure suffer from various types of corrosion: aquifer wells with their extraction pumps and steel casings, steel and concrete pipelines for long-distance water conveying, municipal pipes for potable water distribution and rainwater collection, and all sorts of metallic and plastic tubes and valves for water supply in residential and public buildings. Water systems require the application of corrosion methods and techniques of the facilities. Practical methods that minimise or eliminate corrosion include selection of resistant materials of construction, application of coatings and linings, cathodic protection, and use of corrosion inhibitors.

Several cases of corrosion management, including the application of 'green' corrosion inhibitors, are shown, illustrated, and discussed.

Keywords: inhibitors, corrosion, water, pipelines, steel

19.2 Introduction

Corrosion affects the durability of civil infrastructure assets, including water production, supply, and storage systems. 'Green' corrosion inhibitors will extend the life of industrial equipment for water systems. Special green inhibitors are obtained from plants growing in desert regions of the state of Baja California, Mexico, by ethanolic and aqueous extraction.

19.3 Water Systems

Fresh water comes from rain and snow; it accumulates in rivers and lakes and generally contains <1,000 mg of dissolved solids per litre (mg/L). Potable and building water include low levels of total dissolved solids (TDS), and some chemicals (e.g., chlorine) are added for health reasons. Many types of water are produced, transported, and used: potable for municipal systems, irrigation for agricultural crops, and cooling for industrial facilities, facilities using fossil fuels, and nuclear energy. Figure 1 depicts a plant for treatment and clarification of water for human consumption.

Figure 1. A plant for water treatment and clarification.

Water is conveyed by a pipeline system, which consists of a large number of pipes, pump stations, and valves, that moves the water from a source to the consumption location. Pumps are essential components of water supply systems (figure 2).

Figure 2. Pumps for water transportation.

Usually, water pipelines are fabricated from ductile iron (DI) and from carbon steel (CS) based on the American Petroleum Institute (API) standards, but they may also be constructed from concrete or plastics, including reinforced plastics.[4] A system of painted steel water pipelines is shown in figure 3.

Figure 3. Painted steel water pipelines.

Water pipes have an inner diameter between 0.10 and 2.0 m, and the water flows at speeds of 1 to 6 m/s. Modern water pipelines are operated remotely from computerised control rooms, and satellite surveillance is used to detect leaks or mechanical failures. The water quality and its influence on human health depends on the pipeline performance and whether it is free from corrosion, scaling, and fouling.

Steel corrosion is an electrochemical process that occurs on a pipe surface upon reaction with the water components, mainly dissolved oxygen (DO) and salts. Waters with a high concentration of dissolved and suspended solids, such as carbonates, silicates, phosphates, and hydroxides, form thick scales that might plug the pipes and interfere with water flow. Sometimes macro- and microorganisms thrive in water and cause corrosion.

The water corrosivity is determined by laboratory corrosion tests, simulating industrial conditions, and applying ASTM standards[5] and NACE TM0169.

19.4 Corrosion Protection and Control

The water infrastructure requires the application of corrosion control methods and techniques from the early stages of design through the construction and operation of the equipment. Practical methods that minimise or eliminate corrosion include the selection of corrosion resistant construction materials, application of coatings and linings, cathodic protection (CP), and use of corrosion inhibitors. The most direct means of preventing corrosion is the choice of suitable materials. The final selection, particularly for water pumps, must be a compromise between technological and economic factors.[7]

The purpose of a coating or lining is to act as a non-reactive barrier between the water and the material to be protected, generally steel (figure 4). Coatings fall into three main groups based on their chemical nature: metallic, organic (including paints), and inorganic. CP is based on the electrical nature of corrosion and is usually applied to water pipelines.

Figure 4. Coatings for protection against corrosion pipelines.

19.5 Corrosion Inhibitors

Corrosion can be controlled by modifying the water environment and by neutralising or removing corrosive agents (e.g., DO). Corrosion inhibitors slow the rate of corrosion reactions when added in relatively small amounts to the water.

They are divided into three groups:

- anodic inhibitors, which retard the anodic corrosion reactions by forming passive films
- cathodic inhibitors, which repress the corrosion reaction such as reducing DO
- adsorption inhibitors, such as amines, oils, and waxes, which are adsorbed on the steel surface to form a thin protective film that prevents metal dissolution

These conventional inhibitors are applied in many sectors of the water and energy industries—cooling water systems,[3,8-9] desalination plants,[10-12] coal water slurries,[3] acid pickling of metals,[3] on reinforcing steel in concrete, and for control of galvanic corrosion in heat exchangers exposed to reverse osmosis water (table 1).[12]

In the last decade, a new family of inhibitors has emerged called green corrosion inhibitors, which are relevant in this crucial time of energy problems and economic havoc since they will extend the life of the water infrastructure and save large expenses in materials, equipment, and structures. They belong to the advanced field of green chemistry, also known as sustainable chemistry, which involves the design of chemical products and processes that reduce or eliminate the use or generation of hazardous substances.

In one example, R. Garcia[13] evaluated the inhibitive action of an ethanol extract derived from a desert plant on the corrosion of CS in hydrochloric acid (HCl) and found it to be effective. HCl is employed at times to remove carbonate scales from steel surfaces.[13-14]

Table 1. Applied corrosion inhibitors for water systems (A)

System	Corrosion inhibitors
Engine coolants	• molybdate
	• molybdate with nitrite; molybdate, arsenite, or arsenate and benzotriazole along with borate/phosphate/amine
	• Nitrite, nitrate, phosphate, borate, silicate, benzoate, aminophosphonate, phosphinopolycarboxylate, polyacrylate, hydroxybenzoate, phtalate, adipate, benzotrizole, tolyltriazole, mercaptobenzothiazole, and triethanolamine are combined with molybdate. In glycol, 0.1 to 0.6 wt% of molydbate is used.
Closed recirculating cooling water	• 200 ppm sodium molybdate with 100 ppm of sodium nitrite
	• 50 ppm molybdate, 50 ppm phosphate, 2 ppm Zn^{2+} • 40 ppm sodium molybdate + 40 ppm sodium silicate
	• 2-phosphonobutane-1,2,4-tricarboxilic acid and polyvinylpyrrolidone
Cooling water of steam plant boiler waters	• molybdate with an aluminum salt and thiourea
	• mild steel corrosion inhibition in boilers by a mixture of sodium molybdate, sodium citrate, manganese sulphate, polymaleic acid, and morpholine
	• protection of mild steel in hard water boilers by sodium molybdate and sodium nitrite
(A)Source: V. S. Sastri, Green Corrosion Inhibitors: Theory and Practice, Wiley and Sons (2011).	

19.6 Conclusions

Corrosion is a damaging process that affects the water infrastructure including pipelines, pumps, valves, and auxiliary equipment. Economic considerations are of the utmost importance when evaluating anticorrosion methods involving inhibitors.

The use of an inhibitor under operating conditions is determined largely by both its chemical stability and its corrosion-prevention efficiency. Conventional corrosion inhibitors, especially green corrosion inhibitors, will contribute to maintain effective water systems and their related natural and man-made environments.

19.7 References

[1]   B. Valdez, M. Schorr, et al., 'Effect of Climate Change on the Durability of Engineering Materials in Hydraulic Infrastructure: An Overview,' Corr. Eng. Sci. and Technol. 45, 1 (2010): pp. 34–41.

[2]   B. Valdez, M Schorr, eds., 'Special Issue: Relationship of Corrosion with Climate Change,' Corr. Eng. Sci. Technol. 45 (2010).

[3]   V. S. Sastri, Green Corrosion Inhibitors: Theory and Practice (Hoboken, NJ: John Wiley and Sons, 2011), pp. 212, 216, 223.

[4]   W. Sung, 'Corrosion in Potable Water Distribution and Building Systems,' S. D. Cramer, B. S. Covino Jr., eds., Corrosion: Environments and Industries, ASM Handbook, Vol. 13C, (Materials Park, OH: ASM International, 2006), pp. 8–11.

[5]   Corrosion of Metals, Wear and Erosion, Annual Book of ASTM Standards, Vol. 03.02 (West Conshohocken, PA: ASTM, 2012).

[6]   NACE TM0169-2000, 'Laboratory Corrosion Testing of Metals' (Houston, TX: NACE International, 2012).

[7]   P. Dupont, J. P. Peri, 'World-Class Water Pumps,' Sulzer Technical Review 3 (2011): pp. 12–15.

[8]   B. P. Boffardi, 'Corrosion Inhibitors in the Water Treatment Industry,' S. D. Cramer, B. S. Covino Jr., eds., Corrosion: Fundamentals, Testing and Protection, ASM Handbook, volume 13A (Materials Park, OH: ASM International, 2003).

[9]   A. Abulkibash, et al., 'Corrosion Inhibition of Steel in Cooling Water System by 2-Phosphonobutane-1,2,4-Tricarboxilic Acid and Polivinylpyrrolidone,' The Arabian J. for Sci. and Eng. 33 (1A), 1 (2008): pp. 29–40.

[10]   M. Schorr, B. Valdez, J. Ocampo, A. Eliezer, 'Corrosion Control in the Desalination Industry,' M. Schorr, ed., Desalination, Trends and Technologies (New York, NY: Intech, 2011).

[11]   M. Schorr, B. Valdez, J. Ocampo, A. Eliezer, 'Materials and Corrosion Control in Desalination Plants,' MP 51, 5 (2012): pp. 56–60.

[12]   Carrillo, B. Valdez, M. Schorr, R. Zlatev, 'Inorganic Inhibitors Mixture for Control of Galvanic Corrosion of Metals Cleaning Process Industry,' CORROSION 2012 (Houston, TX: NACE, 2012).

[13]   R. Garcia, B. Valdez, R. Kharshan, A. Furman, M. Schorr, 'Interesting Behaviour of Pachycormus Discolor Leaves Ethanol Extract as a Corrosion Inhibitor of Steel in 1 M HCl: A Preliminary Study,' Intl. J. of Corrosion (2012).

CHAPTER

Corrosion of Copper Coils in Air Conditioning Equipment

A. Calderas, N. Santillan, B. Valdez, and M. Schorr
University of Baja California, Mexico

Atmospheric corrosion affects the brazed unions of copper tubing in heating, ventilation, air conditioning, and refrigeration equipment, causing maintenance problems. As a result, refrigerant gases are emitted directly into the atmosphere and adversely affect the environment. The corrosion behaviour of test coupons made of copper joined with different silver brazing alloys was investigated, and the results are presented.

Corrosion of brazed joints in the components of heating, ventilating, air conditioning, and refrigeration (HVAC/R) installations generates failures of metallic materials and permits the release of refrigerant gases, contributing to atmospheric pollution.[1] The effects of gas emissions on climate change and global warming represent a worldwide concern. The city of Mexicali, Baja California, Mexico, is located in a semi-arid region on the Mexico-United States border. It has an extremely hot climate with temperatures that rise above 45°C during the spring and summer seasons[2] and is classified as one of the most contaminated cities in Mexico.

Different types and sizes of HVAC/R systems have been installed for comfort in residential, institutional, and industrial buildings. Corrosion and cracking of silver brazing alloys (table 1) used to join copper tubing is a common cause of failure, enhanced by the presence of contaminants such as sulphides, chlorides, CO_x, NO_x, and SO_x contained in the urban-industrial atmosphere of Mexicali. To study these phenomena, copper tubing coupons joined with different silver brazing alloys with 0, 5, and 15 per cent Ag were tested over two years by their exposure to real city environments in four different locations with specific characteristics regarding the pollution level.

The conventional air HVAC/R systems used to control temperature, relative humidity, and environmental quality are mainly composed of a compressor, evaporator, condenser, and expansion valve (figure 1). Corrosion occurs on the components of HVAC/R systems, and it increases with time.

20.1 Experimental Procedures

In the HVAC/R technology, copper type L is the preferred material used in the manufacturing of HVAC/R unit coils as well as other tubing in the system because of its excellent thermal conductivity, durable properties, and resistance to corrosion.[3] Sturdy brazed connections of the refrigerant lines, copper coils, and tubes can be prepared employing different silver alloy filler materials (table 1).[3-4]

Table 1. Elements in brazing filler metals[4]

AWS Classification[A]	Principal elements (% w/w)					
	Silver (Ag)	Phosphorus (P)	Zinc (Zn)	Cadmium (Cd)	Tin (Sn)	Copper (Cu)
BCup-2	—	7.00-7.5	—	—	—	Remainder
BCup-3	4.8-5.2	5.8-6.2	—	—	—	Remainder
BCup-4	5.8-6.2	7.0-7.5	—	—	—	Remainder
BCup-5	14.5-15.5	4.8-5.2	—	—	—	Remainder
BAg-1	44-46	—	14-18	23-252	—	14-16
BAg-2	34-36	—	19-23	17-192	—	25-27
BAg-5	44-46	—	23-27	—	—	29-31

[A]ANSI/AWS A5.8 Specifications for Filler Metals for Brazing

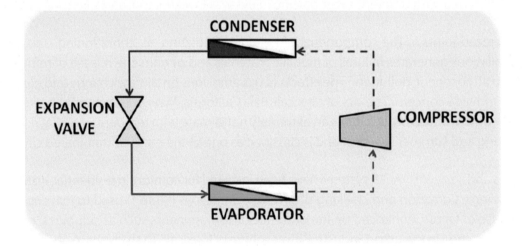

Figure 1. Diagram of the main components of HVAC/R equipment.

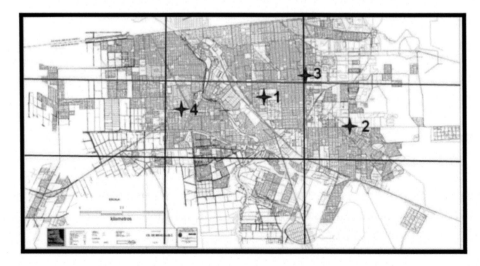

Figure 2. Mexicali map showing corrosion tests locations.

(a)   (b)

Figure 3. Details of the installation of test coupons: (a) annually operated unit, (b) temporal operated unit.

20.2 Corrosion Testing

To analyse the reason for the appearance of corrosion, four Mexicali locations were selected based on prior considerations of the major occurrence of emissions (figure 2). These locations were classified as follows:

- domestic area
- Mexicali wastewater oxidation pond
- proximity to Cerro Prieto geothermal plant
- an industrial area

Copper coupons cut from pipes and joined with silver filler metal were prepared. To be more accurate on the corrosion tests, a 25 mm inside diameter (ID) type L tube was first cut in half and then laminated and cut again to expose a 50 mm^2 surface. The surfaces were polished with fine grit sandpaper and afterwards joined with different B-Cu-P brazing alloy metals (0, 5, and 15 per cent Ag) to make the final coupons with a 100 mm^2 surface (figure 3). The coupons were cooled and then submersed in a hydrofluoric acid (HF) solution for three minutes, rinsed with acetone, and finally dried prior to installation. The practice recommended for corrosion testing in ASTM standards G1,[5] G4,[6] and G31[7] were followed.

A total of twenty-four coupons of each type of brazing material were installed in situ directly on the HVAC/R units (figure 3); twelve of these on units work yearly, and twelve units work only in the spring and summer seasons at each city location. The coupons were positioned horizontally with a distance of 50 mm between them.

Table 2. Estimation of corrosion results

	Unit of operation					
	Annual (AU) silver percentage (%)			Temporal (TU) silver percentage (%)		
Location[A]	0%	5%	15%	0%	5%	15%
1	+	+	+	+	+	+
2	+	+	+	+	+	+
3	+	+	+	+	+	+
4	+	–	–	+	+	+

[A]+ Corrosion – No corrosion (locations are identified on Mexicali City map [figure 2])

(a) (b)

Figure 4. Coupons installed on the second point of study: (a)
annually operated unit, (b) temporal operated unit.

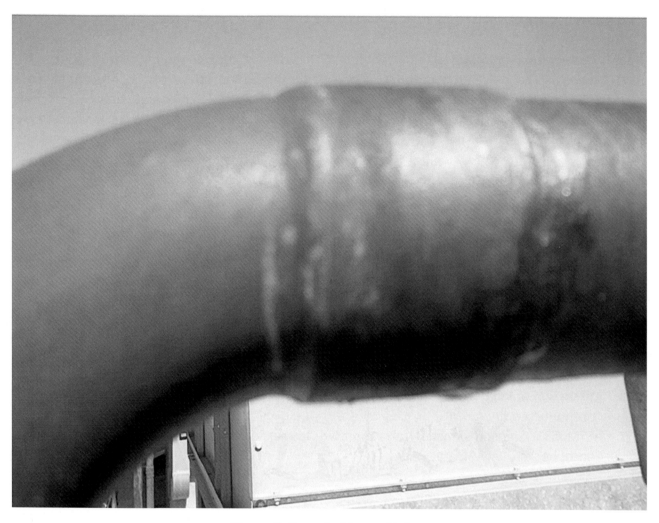

Figure 5. Corroded brazed joints.

20.3 Results

The aggressive and corrosive surroundings at the city location were significant. As a result of this exposure, some coupons underwent a change in colour because of the appearance of stains, revealing that a corrosion process had started to take place. Both localised pitting and general corrosion were detected on the surfaces of copper and the brazed zone after four months of exposure to the environment at the different locations. According to these results, the coupons were classified as units working annually (year-round) or temporarily by seasons: Annual Unit (AU) or Temporal Unit (TU) (table 2).

Figure 4(a) presents details of the change in colour that occurred in units working during the entire year. Tarnishing appears mainly on the brazing, and some loss of material is also present on the brazed surface of the coupon. Figure 4(b) corresponds to units that operate only during the summer season at the second point of the study; here, the blackening appears on the entire coupon, and a well-defined pit is present on the right side of the brazing. Some pits are also visible on the copper base material. Some brazed connections were corroded (figure 5).

20.4 Conclusions

The polluted environment plays an important role in the cause of corrosion in the brazed joints of the HVAC/R tubes and coils. Even though the coupons were exposed for a short period, relevant changes occurred. The blackening of the coupons installed on the HVAC/R units that operate only by seasons is more remarkable (figure 4[b]). In all study locations, the brazing without silver exhibits a greater tendency to corrode than those with 5 or 15 per cent Ag. In addition to a change in coloration, they exhibited several pits.

20.5 References

1. B. Valdez, et al., 'Effect of Climate Change on Durability of Engineering Materials in Hydraulic Infrastructure: An Overview,' Corros. Eng. Sci. and Technol. 45, 1 (2010): pp. 70–75.
2. R. García, N. Santillán, 'Modelling Extreme Climate Events: Two Case Studies in Mexico Climate Models,' Climate Models (Rijeka, Croatia), ISBN: 978-953-51-0135-2 (2012), pp. 145.
3. L. Jeffus, 'Handbook of Air Conditioning and Refrigeration,' 4th ed. (Upper Saddle River, NJ: Prentice Hall, 2004): pp. 114–127, 136–140, 194–209, 510–511.
4. ANSI/AWS A5.8, 'Specification for Filler Metals for Brazing' (Miami, FL: AWS, 2011).
5. ASTM G1, 'Standard Practice for Preparing, Cleaning, and Evaluating Corrosion Test Specimens' (West Conshohocken, PA: ASTM).
6. ASTM G4, 'Standard Guide for Conducting Corrosion Coupon Tests in Field Applications' (West Conshohocken, PA: ASTM).
7. ASTM G31, 'Standard Practice for Laboratory Immersion Corrosion Testing of Metals' (West Conshohocken, PA: ASTM).

CHAPTER 21

Corrosion Assessment of Infrastructure Assets in Coastal Seas

Benjamín Valdez,[a] Jorge Ramirez,[b] Amir Eliezer,[c] Michael Schorr,[a] Rogelio Ramos,[a] and Ricardo Salinas[a]

[a]Laboratorio de Corrosión y Materiales Avanzados, Instituto de Ingeniería, Universidad Autónoma de Baja California, CP 21280 Mexicali, México.
[b]Ciencias de la Tierra, Instituto de Ingenieria, Universidad Autónoma de Baja California, CP 21280 Mexicali, México.
[c]Corrosion Research Centre, Sami Shamoon College of Engineering, Beer Sheva, Israel

21.1 Abstract

The seas play an essential role for the peoples living on their coastal regions, since the marine infrastructure is located in the coasts. Seawater is a corrosive environment that affects infrastructure, particularly in polluted seawater. Corrosion and pollution are pernicious chemical, physical processes that impair the quality of the environment and the durability of the marine structures and materials. They are aggravated by the discharge into the sea coast of municipal, industrial, and agricultural effluents, which contain and produce toxic and highly corrosive components by biological and chemical degradation. Reinforced concrete and carbon steel are the main engineering materials used for the construction of marine installations and equipment, but other metals and alloys—aluminium, copper, stainless steels—are applied, too. Laboratory and field corrosion tests in seawater were carried out applying gravimetric, electrochemical and surface examination methods, based on American Society of Testing and Materials (ASTM) and National Association of Corrosion Engineers (NACE) standards. This work is the result of a cooperation between academic institutions in Mexico and Israel. The data generated advance the management of sea corrosion prevention and mitigation and provide a guide for marine infrastructure maintenance and corrosion control. Several cases of corrosion in the sea coasts based on the authors' experience and knowledge are presented.

21.2 Introduction

The seas played an essential role for the humans living on the coastal regions of the world. These ancient maritime civilisations were widespread by peoples of the Mediterranean, Arabian, and China Seas and the Atlantic and Pacific Oceans, managing complex systems of harbours, shipyards, and seafaring for trade, conquest, and colonising expeditions. The maritime heritage of the Phoenicians, Greeks, Romans, Vikings, Byzantines, and Spaniard navigators has been implemented through the millennia by the modern maritime nations (Leon H. Charney School of Marine Sciences 2016).

The coast and its adjacent areas onshore and offshore are an important part of local ecosystems, forming gulfs, bay, and estuaries, sometimes mixing fresh and salty waters.

The seas energy is enormous, tides rise and fall twice daily, and swift currents flow towards the coast. Offshore electricity generators and waterfront industries such as liquefied natural gas (LNG) regasification plants, with their cryogenic ships and mooring ports, will power coastal and inland cities (Valdez et al. 2011).

The Gulf of Mexico (GOM), an extension of the Atlantic Ocean, characterised by its tropical climate, reaching the coasts of Mexico and US, displaying elements of infrastructure, was evaluated in the framework of this work. The GOM was chosen as a typical sea coast of two countries seriously affected by corrosion and pollution (Hernandez et al. 1995; Acuña et al. 2002).

Additional gulfs and areas with infrastructure assets and environmental characteristics similar to those encountered in the GOM exist in the worldwide tropical zone—for example, Gulf of California, of Panama, of Suez, of Aqaba, the Arabian, the Persian, the Bengal Bay, and the Strait of Singapore. The experience, knowledge, and information gathered in this work will contribute to corrosion control and preservation of the infrastructure in these marine tropical regions.

The aim of this investigation is to build a body of knowledge useful for the selection of corrosion-resistant engineering materials for materials for the marine infrastructure and naval industry.

21.3 Corrosion and Pollution Interaction

Corrosion and pollution are interrelated processes since many pollutants produced by power stations burning fossil fuels accelerate corrosion, and corrosion products such as rust, oxides, and salts pollute water bodies. Both are pernicious processes that impair the quality of the environment and the durability of the marine structures and construction materials (Raichev et al. 2009).

They are permanent problems in the fluvial and marine environments, in maritime activities and naval industries. This situation is aggravated when industrial, municipal, and agricultural pollutants flow into the river, the port, and its surroundings, damaging its structures, installations, equipment, and machinery, The physical, chemical, biological, and thermal characteristics of these effluents affect the entire sea coast environment and influence the corrosion processes, types, and mechanisms. These natural and anthropogenic pollutants increase the corrosion extent of steel, reinforced concrete, and other engineering materials of the infrastructure assets. The knowledge of their source and characteristics contributes to the establishment and management of technologies to control water pollution and industrial corrosion, applying corrosion monitoring and corrosion risk assessment. Furthermore, water pollutants, even when present at very low levels, impair human health, aquatic life, and seawater quality. The avoidance of pollution is therefore an important part of corrosion prevention and control (Marcos et al. 2006).

21.4 Climate Change

There is nowadays a deep universal concern about the influence of climate change (CC), global warming, and greenhouse emissions (all interrelated complex phenomena) on the corrosive and deterioration effects of the marine environment on maritime activities. Higher temperatures promote melting of polar and glacial ice and snow; sea level rise affects the coast structures and changes atmospheric moisture and rain precipitation patterns. The climate and extreme weather-related events (such as floodings, strong winds and storms coming from the seas that erode the coasts, prolonged heat waves, dust and particulate matter flowing in the air) seriously affect the environment and society in a variety of ways, acting through destruction of the infrastructure, famine, disease, and social and economic upheaval. But the powerful floods that ripped roads and bridges are the most devastating events wreaking havoc across vast regions with great loss of life and property.

This growing concern has been expressed by IMarEST, the Institute of Marine Engineering, Science and Technology, UK, which recognise that CC is the most serious threat facing humanity, resulting in significant warming of the oceans and the marine atmosphere. Furthermore, the Institute of Materials, Minerals and Mining (IOM3), London, published a special issue of its journal, *Corrosion Engineering, Science and Technology*, entitled 'Relationship of Corrosion with Warming and Climate Change', which brings together papers examining CC-induced corrosion (Valdez et al. 2010). Roberge (2010) reports on three aspects of the effect of CC: corrosivity at coastal regions, increased stress on marine systems, and pluvial precipitation pattern that may change the corrosive behaviour of the environment and increase the risk of corrosion failures. Measurements are presented of coastal erosion and corrosion and flooding of the British coasts. Corrosivity maps for 2000 and 2100 illustrate the present and future situation (Roberge 2010).

21.5 Seas and Rivers

Human civilisation begins and spreads along the riverbanks; when the population increase, cities are created and ports built. The major cities of Europe and America were installed on the banks of the Danube, Thames, Guadalquivir, Tiber, Mississippi, Parana, which flow into deltas, estuaries, and, finally, into the seas of their regions. The dominant presence of water affects the stability of materials and engineering structures by physicochemical processes of absorption, penetration, dissolution, hydration, hydrolysis, oxidation, carbonation, erosion, corrosion, and degradation; therefore, modern technologies of prevention, protection, and control are applied (Marcos et al. 2006).

The river flow is not constant; melting snow and heavy rains increase its flow, causing flooding and damaging turbulence, which cover vast expanses of land, degrading and destroying buildings, bridges, roads, structures, and facilities.

The sea is a dynamic system in permanent motion, with complex surface currents and winds blowing over its surface, generating waves that reach the coast, its facilities and installations. In splash zones, violent, rough waves break down, generating whitish oxygenated foam, hence increasing local corrosion. Sometimes, these high, powerful waves erode and devastate the coast.

Seawater consists of a solution of many salts and numerous organic and inorganic particles in suspension. Its main characteristics are salinity and chlorinity and, from the corrosion point of view, dissolved oxygen (DO) content, which ranges from 4 to 8 mg/l, depending on temperature and depth. Its minor components include dissolved gases—CO_2, NH_3, and H_2S—found in seawater contaminated by urban sewage. The oceans house algae, bacteria, and phytoplankton, which generate about half of the oxygen in the atmosphere (Melcher 2005).

Ocean surface salinity is determined by the balance between water lost by evaporation and water gained by precipitation. The salt concentration, particularly NaCl, varies from 3.5 per cent to 2.0 per cent, according to the sea location and the massive addition of fresh river water. For instance, in the Red Sea (an enclosed basin), salinity at high summer temperatures is 4.1 per cent; but in the Baltic Sea, it is about 2.0 per cent since many rivers feed into it.

The expansion of industry, transportation, and the energy sector increase the CO_2 air content. This CO_2 dissolves in the seas, decreasing their pH from 8.11 to 8.06, acidifying with carbonic acid (H_2CO_3), particularly near the coast where the fossil fuel power stations are located. Acidic water affects the life of marine calcifying organisms such as coral, molluscs, and plankton (Bonilla et al. 2011; Anonymous 2013).

21.6 Marine Infrastructure

The infrastructure of a nation constitutes the physical base of its economic activity; its quality represents a crucial index of the industrial and social vitality of a country. During processes of growth and development, national governments emphasise the maintenance and modernisation of the infrastructure to assure the long life of these assets. Budgets and huge efforts are spent to operate, maintain, and upgrade the coastal seas infrastructure, which are composed of two types of structures: fixed and mobile. Fixed structures are on the gulfs' and bays' commercial and industrial ports, shipyards, rail yards, fuel terminal, naval bases, platforms for oil drilling, production, processing, and storage; submarine pipelines, and communication cables and electricity grid lines along the coasts (Valdez et al. 2015).

The mobile structures include all types of ships, large and small: civil, commercial, and military. These include general bulk cargo vessels for cereals, chemicals, minerals, cryogenic vessels for LNG, cruise tourist and recreational yachts, fishing vessels, container ships, oil tankers, military ships including cruisers, aircraft carriers, patrol boats, light, nuclear and conventional submarines. The International Maritime Organization (IMO), a United Nations Agency, estimates that at any time about one hundred thousand ships and vessels roam through the oceans and seas of the world!

The infrastructure elements are located along the coastline. Many infrastructure assets (roads, railways, bridges) were not planned or constructed to handle the actual traffic loads and need repair and rehabilitation work. There are seventy thousand structurally deficient US bridges, and more than 15 per cent are at risk of catastrophic corrosion-related failures (Hummel 2014).

Table 1 presents a summary of the elements of infrastructure existing on the sea coast and the adjoining mainland. They are made of a great variety of materials of construction: steels, SS, reinforced concrete, plastics and composites, elastomers for lining of storage tanks and pipelines. The correct selection of construction materials will assure the infrastructure resiliency in case of weather extremes.

21.7 Ports, Shipyards, and Ships

Ports for maritime commerce and shipyards for the construction of wooden seagoing sailing vessels were established in ancient times along the coasts of the Mediterranean Sea by the Phoenicians, the Greeks, and the Romans.

Today, a marine port situated near a large city belongs to a coastal ecosystem that comprises industrial parks, agricultural fields, and a hydrologic basin with streams that flow into the sea in the vicinity of the port. Modern ports are the central link between maritime and land transportation, responsible for the exports and imports that ensure the prosperity of a nation. The port's access, with its dredged channels to maintain a sufficient depth, allows the manoeuvring of ships. This active traffic depends on the port installations and equipment, their quays, wharves and docks, tugs, internal trains, and warehouses. Breakwaters, which protect the port from powerful waves, are built entirely of large stones or rocks, or seawater-resistant concrete boulders or tetrapods. With the growth of trade, commerce, and tourism and the increase in the size of oil tankers, passenger and cargo ships, many ports are being enlarged, their dock accommodations extended, and their entry channels deepened. This modernisation of port facilities and equipment is a national priority and is being actively implemented.

Table 1. Sea coasts infrastructure

Infrastructure	Assets	Corrosion problems, protection, and maintenance
Energy generation	Thermoelectric, nucleoelectric, eolic and hydroelectric plants, dams, electricity grids: cables and towers	Hot corrosion by fossil fuel combustion gases control of fuel sulphur and vanadium contamination. Erosion corrosion and stress corrosion cracking in turbines. Feed water treatment
Waterworks	Water supply: pipelines and pump stations, sewage systems water and wastewater treatment plants, pluvial drainage	Control of corrosion, scaling and fouling by correct design. Selection of materials and coatings. Cathodic protection for steel pipelines. Coatings for concrete and steel pipelines
Petroleum industry	Offshore platforms, production wells, marine oil terminals, submarine and land oil and gas pipelines, refineries. LNG regasification plants	Corrosion by sour and sweet gas and crude. Corrosion control with inhibitors. Cathodic protection for steel platforms and pipelines. Control of corrosive emissions
Transportation	Harbours, shipyards, airports, river, railway and seawater bridges, railways, highways and roads, waterways, public transportation	Asphalt pavement deterioration. Repair and restoration of steel-reinforced concrete structures. Use of coatings and cathodic protection. De-icing of roads and bridges deck
Communication	Underground cables: telephone, TV, electricity. Communication towers and antennae	Protection of buildings from atmospheric marine corrosion with water-repellent plastics and paints. Restoration of facades with polymeric concrete
Construction	Housing, public buildings: hospitals, hotels, stadia, urban recreation areas, parking sites	Building protection from weather by paints and coatings. Restoration of facades with polymeric concrete
Fishing	Harbours, marine vessels, freezing and packaging plants	Marine corrosion and fouling. Anticorrosive and antifouling paints and cathodic protection

The shipyards operate two types of docks: a floating one and a dry one. The first serves for inspection, maintenance, and repairs of ships; it incurs in high maintenance costs because the steel structure requires regular removal of corrosion products, sea fouling and scale, and periodical painting to prevent or minimise corrosion. A floating dock consists of a structure with a U-shaped profile, built of steel, with a double wall deck and a high lateral series of ballast chambers, which, upon filling with seawater, causes the dock to become submerged. In this way, the entry of the ship to be repaired is facilitated, and upon emptying the chambers by pumps, the dock emerges from the water with the ship resting on its dry deck. Floating docks incur high maintenance costs because the steel structure, being continuously afloat, requires the regular removal of sea fouling and scale and occasional painting, as the hull of a ship does, to prevent or minimise corrosion. When the water in a port is contaminated and the sea soil beneath the floating dock contains corrosive sediments, accelerated corrosion results in a large number of cavities and perforations in the floating dock floor and the seawater chambers (Schorr and Valdez 2005).

A dry dock is a narrow basin made of concrete, closed by steel gates, used for the building of ships. After the ship is completed, the dock is flooded, and it slides slowly into the sea coast, ready to cross the oceans. The dock is also employed for repair and maintenance of all kinds of watercrafts. Modern merchant and military ships built mainly from steel are sailing the seas and oceans or moored at piers ports; therefore, they are affected by corrosion and fouling.

In biblical times, Noah coated his ark 'within and without with pitch' (Gen. 6:14); the Phoenicians, Greeks, and Romans protected their wooden vessels with natural materials: pitch, tar, wax, sulphur, asphalt, resins, and tallow. The Royal British Navy applied copper sheets to avoid the attachment of fouling organisms. Applying modern marine paints constitute the primary method of corrosion control, with cathodic protection, by impressed current or sacrificial anodes, as a supplementary method. The hull plates of a ship are pre-treated with a phosphate coating before ship assembly at the shipyard, so as to protect them against corrosion (Valdez 2016).

Ancient and modern shipwrecks are investigated by scientists and divers, trained in coastal and underwater archaeology, to discover the corrosion mechanism of metallic artefacts buried in the seafloor for centuries. Artefacts recovered include military suppliers (produced from wrought and cast iron), religious jewellery (made of gold, silver, copper, and bronze), and ship hardware. These activities constitute the base for the study of maritime civilisation.

Nowadays, Europe, Japan, South Korea, and Singapore yards are building large oil barges, oil rigs, and drill ships designed for work at the deepest water, such as in the GOM. These are a new generation of fuel-efficient, cheap-to-run 'Eco ships'. These platforms, and the metallic underwater pipelines used to transfer oil and gas from the borehole to the surface, are fitted with sacrificial anodes for corrosion control by cathodic protection. They are towed out to sea and installed at various depths around the world.

21.8 Ballast Water Technology (BWT)

Huge petroleum tankers haul crude and derivatives from the coast sea terminals to the refineries and consuming markets. They suffer from corrosion in the steel holds; when they cross back to the terminals, the holds are full of ballast seawater, protected with corrosion inhibitors (CIs) (4th Ballast Water Technology Conference 2015). They are added to the ballast seawater as a fine powder, which converts the seawater into a colloidal suspension with particles dispersed in the water that are adsorbed on the steel surface, forming a thin protective film.

A study has been carried out by analysing the water corrosivity, composition, hardness, conductivity, and pH. The mechanism of the formation of a gelatinous white precipitate is based on the combination of the water Ca^{2+} and Mg^{2+} ions with the CI chemical structure (Cheng et al. 2016).

About 4,400 petroleum transportation tankers from oil-producing countries cross the oceans and seas of the world to energy-consuming countries. If, on average, each tanker has 10 cargo holds, it means 44,000 holds require a CI for their ballast seawater.

Petroleum steel tankers (figure 1) are cheaper and more efficient than submarine pipelines installed on the seabed for oil transportation. For their trip back, the tanker holds are filled with seawater to provide adequate stability (figures 2 and 3) and safety.

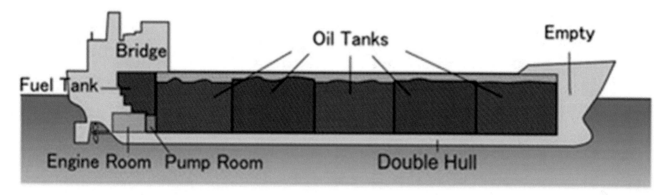

Figure 1. Petroleum transportation tanker showing holds.

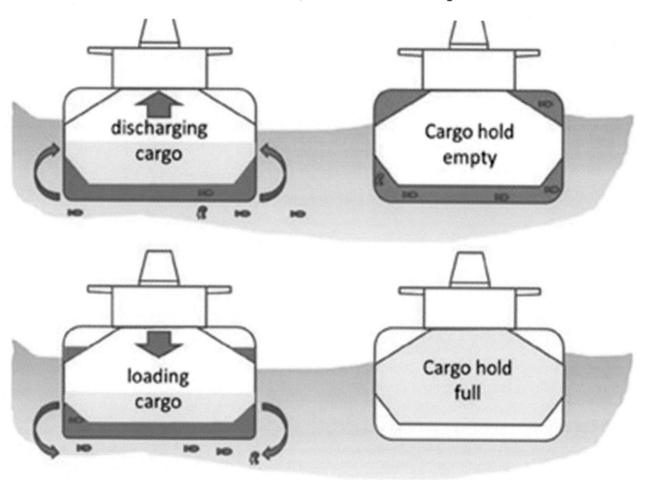

Figure 2. Diagram of ballast water intake and cargo discharge and loading.

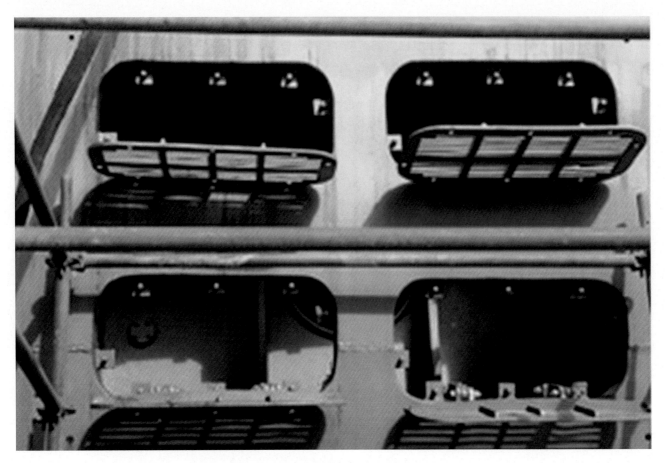

Figure 3. Ship opening, fitted with screens, for ballast water intake and discharge.

Recently, the Fifth IMarEST Ballast Water Technology Conference was held in London to deal with the challenges facing the shipping industry. The two main concerns are the avoidance of the ship holds corrosion and the pollution of the ports and the adjacent sea regions by discharging non-native aquatic species. Two central maritime entities—IMO and the US Coast Guard (USCG)—are involved with the regulations managing these key issues (18th ICMCF 2016).

21.9 Marine Corrosion

In recent decades, the concept of corrosion has been expanded to cover the deterioration in seawater of structures and equipment manufactured from non-metallic materials—for example, polymers and composites—used for the fabrication of recreation boats, naval vessels, and coastguard ships. However, the aqueous corrosion of metallic structures in marine environments is an electrochemical process that occurs on the metal surface by interaction between it and the constituents of saline water. The dominant factors are salinity and the concentration of DO. Salinity influences the conductivity of the water, and the chloride (Cl^-) ions also affect the oxide layer formed on the metal surface. When Cl^- penetrates the passive film, it can initiate pitting and crevice corrosion at localised sites, with breakdown of passivity.

During the corrosion process, marine structures lose wall thickness at a rate that varies with depth. Normal corrosion rates (CRs) for steel in seawater lie in the range 0.1–0.3 mm y^{-1} but can increase to 2–4 mm y^{-1} in seawater contaminated with corrosive effluents. Typical cases of severe corrosion are found in the steel retaining walls that separate the land and water in ports. The walls are formed from steel sheet piling about

26 m in length, 3–5 m wide, and 10–12 mm in thickness, driven into the marine soil. The sheets interlock together with grooves or guides along each edge, producing a unit of great strength and stiffness. The walls lose thickness by corrosion until perforation of the plate eventually occurs. The corrosion is accelerated by the impact of waves, which break the fragile rust layer that is removed and falls on the seabed near the wall. With time, the corroded plates must be repaired or replaced to protect the port.

Steel and aluminium alloys are the main metallic materials utilised for the manufacture of civil, industrial, and military ships. Their limited corrosion resistance means that they need to be protected by paints and coating and by cathodic protection. The corrosion extent is increased in turbulent flow conditions since it destroys the rust or scale barrier and provides more DO. Fouling tends to reduce corrosion attack by restricting access to DO, but corrosion takes place under the fouling by acidic excreta. Polluted seawater greatly accelerates the attack on steel. Naval aluminium vessels are built of the 5XXX alloys series in which magnesium (Mg) is the principal alloying element. UNS A95052 is the main alloy used for marine environments. In steel-hulled ships, the upper structures are constructed with AlMg alloys (Eliezer et al. 2010). Modern navies and coast guards of many nations use vessels built with A95052, which provide long-life service. The underwater hull is the most vulnerable part, but corrosion also appears in the propeller shaft and blades, pumps, engines, and fittings. When a badly corroded propeller breaks down, the boat is soon adrift.

A ship navigating in coastal shallow water and the open sea is exposed to three diverse environments and corrosive conditions: (a) the underwater hull, submerged in seawater, covered with marine fouling and supplied with plentiful oxygen; (b) the deck area exposed to sweeping waves, salt spray and heated by solar radiation; and (c) the upper structure exposed to marine atmospheric corrosion. A similar situation occurs with three levels of corrosion at sea-installed petroleum platforms.

Stainless steels (SS) are an important class of alloys extensively used in marine environments, in particular in the oil and gas industry, because they combine high mechanical properties and satisfactory corrosion resistance. They are utilised in the offshore petroleum as platforms, serving in areas as such drilling, subsea piping, pumps valves, flanges, and fittings. Austenitic and duplex SS are required for high-temperature tubulars and raiser pipes.

An International Congress on Marine Corrosion and Fouling has been held recently in France, with IMarEST as one of the sponsors. It provided an interdisciplinary programme that highlights research efforts understanding biofouling and corrosion of materials and structures immersed in the marine environment. The congress dealt not only with the basic sciences of sea chemistry, biology, microbiology, paints coatings but also with real problems of marine structures such as biofouling, adhesive strength, biocides, inhibitors, and protection.

21.10 Marine Biofouling

The corrosion of metallic surfaces such as the ship hull in permanent contact with seawater is caused by the accumulation of macro- and microorganisms: algae, molluscs, barnacles, zebra mussels, bacteria, etc. Their adhesion is mediated by the secretion of glue-like substances of extracellular polymers. Corrosion occurs by anodic and cathodic electrochemical reactions promoted by the corrosive metabolites produced by the biofouling fauna and flora (Railkin 2004).

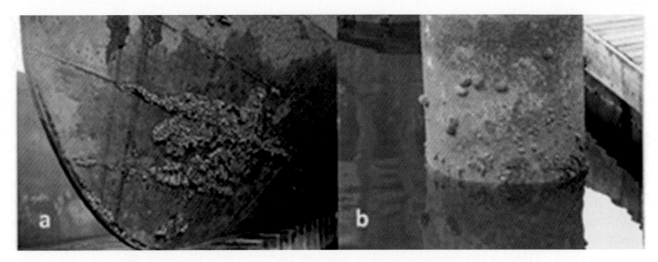

Figure 4. Biofouling cover, (a) ship hall, (b) steel pile for mooring boats.

Biofouling on marine vessels can raise the hydrodynamic volume of the vessel, augment the frictional effects loading to an increased drag, reduce the ship speed, and rise its fuel consumption.

Many fluvial and marine structures are affected: ports, steel retention wall, shipyards, buoys, oil and gas platforms, coaling water towers of power plants. The physicochemical properties of the marine structure's surfaces such as roughness, texture, construction material, wettability, interfacial alkalinity affect the settlement of marine organisms during biofouling community development (Schultz et al. 2011; Huang et al. 2016) (figure 4).

The control of marine biofouling is realised, essentially by application of paints and coatings containing special biocides that are liberated when they come into contact with seawater, preventing the attachment of the marine fouling organisms. Non-toxic antifouling coatings, which contain polymers with antimicrobial activity, have been developed and applied (Hellio and Yebra 2009).

The economic importance of the antibiofouling activities, in particular in the shipping industry, is demonstrated by the involvement of national and international organisations that regulate the industry such as International Maritime Organization (IMO), Integrated Pollution and Control (European Commission), International Antifouling System (IAFS), International Chamber of Shipping, International Convention for the Control and Management of Ships Ballast Water and Sediments, International Council for the Exploration of the Sea, and more.

21.1 Influence of Pollutants

The quality of the river water that flows into a seaport and its surroundings depends on its physical, chemical, biological, and thermal characteristics. The pollutants arrive with the industrial, agricultural, and municipal wastes generated in the region. The industrial plants located around the river produce agrochemicals, petrochemicals, industrial acids (sulphuric, nitric, phosphoric, etc.), plastics, food and beverages, electronic equipment, petroleum distillates, and lubricants. They supply soluble pollutants such as acids and salts, insoluble inorganic minerals, and heavy metal compounds. During periods of intense irrigation or copious rain, fields sloping towards the river that are used for agricultural purposes contribute pollutants containing biodegradable organic matter and nutrients (phosphorous, nitrogen, and potash) from the synthetic and organic fertilisers dispersed thereon. In addition, numerous meat and seafood packaging and cold storage plants; dairy, poultry, and pig farms sometimes pour waste loaded with organic compounds into the river. Biological and chemical processes convert this matter into toxic and corrosive substances. For example, hydrogen sulphide (H2S), with an unpleasant smell of rotten eggs, is biologically oxidised and converted into sulphuric acid, which is extremely corrosive towards steel and reinforced concrete (Schorr et al. 2006):

$$H_2S + 2O_2 \rightarrow H_2SO_4$$

Power plants, producing electricity by the combustion of fossil fuels, cool their condensers with seawater or river water. The hot cooling water is then discharged into the river, accelerating the chemical and biological processes of decomposition and putrefaction, producing methane (CH_4) and ammonia (NH_3) gases. Effluents with suspended solid particles that float towards the port and the coast are slowly deposited on the seabed, forming a thick mud composed of silt, clay, sand, alumina, silicates, organic matter (carbohydrates, proteins and fats), and mineral components: oxides, hydroxides, and compounds of heavy metals. These settled pollutants may persist for many years, so they have to be removed periodically by dredging, and the soil is remediated and restored. They contain toxic and corrosive components that annihilate the aquatic, vegetal, and animal life, thus enlarging the quantity of organic matter and increasing its biological decomposition with the production of H_2S, a corrosive agent. The smell of the decomposition products constitutes an early environmental alert of the probable risks and damage to the port infrastructure. The range of pollutants in typical sea coast or river water receiving industrial, agricultural, and municipal wastes is presented in table 2. The concentration varies with the season, type of industry, agricultural activity, and the effluents discharged.

Table 2. Concentration of pollutants for typical river water and sea coast

Pollutant	Range(mg/l)
Total solids dissolved and suspended	350–1200
BOD (biological oxygen demand)	100–500
COD (chemical oxygen demand)	200–1000
TOC (total organic carbon)	100–400
Nitrogen (N, total)	50–100
Ammonia (NH3)	20–100
Phosphor (P, total)	100–1000
Organic sulphide (S)	50–100
Fat and oil	50–100
Heavy metals	100–1000

21.12 Results

The corrosive substances that affect the fluvial and marine coastal infrastructure assets originate directly from the natural environment, from the industrial effluents formed in situ by biological and chemical conversion of the organic matter supplied by the agricultural and municipal wastes (table 2).

21.13 Corrosion Testing

Laboratory and field corrosion tests in seawater and/or artificial seawater were carried out applying gravimetric, electrochemical, and surface examination methods, based on ASTM, NACE appropriate standards (ASTM 1963; NACE 2000). The practices recommended in the following standards were followed:

- G1, preparing, cleaning, and evaluating corrosion test specimens
- G3, electrochemical measurements in corrosion testing
- G4, conducting corrosion test in field applications
- G5, potentiostatic and potentiodynamic anodic polarisation measurements
- G31, laboratory immersion corrosion testing of metals
- G59, conducting potentiodynamic polarisation resistance measurements
- D1441, preparation of substitute ocean water
- NACE TM 01692000, laboratory corrosion testing of metals

21.14 Metals and Specimens

The diversity of the materials used for shipbuilding is conditioned by the requirements of the ship's structural components, their function, and the ship operating characteristics. The chemical composition of the alloys selected for corrosion testing is listed in table 3.

For the sake of brevity, they are classified into three groups:

- steels, including carbon steels (called hull steels), low alloyed steel, and SS
- copper-based alloys, such as bronzes and brasses, which serve in the ship for water supply and sanitary systems, air conditioning, heat exchangers, propellers, piping and fitting, etc.
- aluminium-based alloys, pertaining to the series 5XXX—for example, UNS A95052 for the manufacture of navy fast vessels and recreation boats.

Table 3. Chemical composition of alloys tested

								Chemical composition											
Materials	UNS	Fe	C	Mn	S	P	Si	Cr	Mo	Ni	N	Cu	Sn	Pb	Al	Zn	Mg	V	Ti
Carbon steel	G10450	Bal.	0.43–0.50	0.60–0.90	0.05	0.04	–	–	–	–	–	–	–	–		–	–	–	–
Stainless steel	S31600	Bal.	0.08	2.0	0.03	0.045	0.75	16.0–18.0	2.00–3.00	10.0–14.0	0.10	–	–	–		–	–	–	–
Copper	C12200	–	–	–	–	0.02	–	–	–	–	–	99.90	–	–		–	–	–	–
Bronze	C52100	0.10	–	–	–	0.30	–	–	–	–	–	Bal.	8.00	0.05		0.20	–	–	–
Brass	C26000	0.05	–	–	–	–	–	–	–	–	–	70	–	0.07	Bal.	–	–	–
Aluminium	A91050	0.4	–	–	–	–	0.25	–	–	–	–	0.05	–	–	Bal.	0.05	0.05	0.05	0.03
Aluminium	A95052	–	–	–	–	–	–	0.25	–	–	–	–	–	–	97.2	–	2.5	–	–

21.15 Corrosion Measurements

Figure 5. The marina at Ensenada port, México.

The corrosion measurements were performed in several marine locations:

- in the GOM coast, at the State of Veracruz, near the mouth of rivers, Mexico
- in the Lerma naval base, a part of the Sound of Campeche, in the state of Campeche, Mexico
- the Ensenada port, on the Pacific Ocean, in the state of Baja California (figure 5); Ensenada port seawater was brought to the Institute of Engineering, at Mexicali, for additional corrosion tests.
- at the Ashdod port on the Mediterranean Sea, done by the Corrosion Research Centre of the Sami Shamoon College of Engineering, Israel

Corrosion tests were performed by two techniques:

- Electrochemical tests based on ASTM standards, G3 and G5, to determine the electrode potential and the corrosion current density, which is expressed as CR in mm/y. The arrangement comprises a polarisation cell containing a three-electrode system: a specimen of the metal being tested (the working electrode), an auxiliary electrode, and a reference electrode, immersed in seawater. A potentiostat was used to obtain the electrochemical parameters.
- A weight loss technique was applied, expressing the loss of wall thickness, in mm/y.

The results presented in tables 4, 5 and 6 are representative of the tests done in several marine locations, as described above.

Table 4 indicates the CRs of the alloys tested in GOM water. The active alloys (e.g., carbon steel, bronze, and brass) show a similar CR in the range 0.10–0.22 mm/y. On the other side, passive alloys display a reduced CR.

Table 4. Corrosion rates of alloys tested in GOM seawater

| Material | UNS | Seawater | | CR(mm/y)[a] |
		type	Location	
Carbon steel	G10450	Natural	Campeche	0.22
Stainless steel	S31600	Natural	Campeche	0.0033
Copper	C12200	Synthetic	UV	0.059
Bronze	C52100	Natural	UV	0.10
Brass	C26000	Synthetic	UV	0.20
Aluminium	A91050	Natural	UV	0.018

[a]CR: Corrosion rate, Sound of Campeche, UV: Universidad Veracruzana, Anticorrosion Unity.

Table 5. Corrosion rates of steel and aluminium in sea and river waters

| Alloy | Water | pH | Corrosion rate[a] | |
			mm/y	Mpy[b]
Carbon steel	Sea	7.5	0.28	11.2
Carbon steel	River	2 to 3	0.82	32.3
A 95052	Sea	7.5	0.05	2.05
A 95052	River	2 to 3	1.2	42.5

[a]Laboratory test by weight loss simulating conditions in the Kishon port and river. [b]mpy = milli inch per year.

Laboratory tests in sea and acidified river water (to simulate the condition of water contaminated by acidic waste effluent) show higher CRs than those of passive alloy A95052 (table 5).

Alloy A95052 exposed in two seas: the Pacific Ocean and the Mediterranean Sea under diverse flow conditions shows a similar open circuit potential but different CRs. The CR measures in the Ashdod port marina is greater than that obtained under laboratory conditions because of the sea dynamic conditions and the plenty supply of DO (table 6).

Table 6. Electrochemical characteristics of UNS a95052 in seawater, at Ensenada, Mexico, and Ashdod, Israel, seaports

Port	Laboratory[a]	Condition	Potential(V)	CR[a](mm/y)
Ashdod Marina	–	Flow	−0.9 to −1.0	1.25 to 1.75
Ensenada	II	Stagnant	−0.8 to −1.2	0.58 to 0.68
Ensenada	II	Flow	−0.8 to −1.2	0.11 to 0.53
Ashdod	CRC	Stagnant	−0.8 to −1.0	0.60 to 0.70
Ashdod	CRC	Flow	−0.9 to −1.0	0.24 to 0.020

ᵃII: Institute of Engineering, University of Baja California, Mexico; CRC: Corrosion Research Centre, Sami Shamoon College of Engineering, Israel; CR: Corrosion rate.

21.16 Conclusions

- The ancient maritime civilisations gradually developed in the sea coasts. Their maritime technologies were adapted and further improved by the modern seafaring nations.
- Combined corrosion and pollution of the marine environment, aggravated by the discharge of industrial, municipal, and agricultural effluents, should be treated by methods of prevention, monitoring, and control.
- Ports, shipyards, and ships are primordial assets of the marine infrastructure that should be protected against corrosion applying paints and coatings and cathodic protection.
- The corrosion resistance value of the metals and alloys tested (listed in table 4) is expressed by the following sequence:

CS < Al alloys < Cu alloys < SS

Local and national environmental organisations such as EPA, in USA and SEMARNAT, in Mexico, in collaboration with ports and rivers authorities and with the assistance of ecologist groups, need to develop and implement regulations covering the prevention of the degradation of their installations, materials, and structures.

To rehabilitate the sea coasts, their rivers, and their surrounding environment, the regional enterprises must stop downloading their effluents, clean the riverbed by dredging, remediate the soil, restore aquatic life, and develop the river landscape for the benefit of the general public.

21.17 Acknowledgements

The authors are grateful to the following institutions and organisations: Petroleos Mexicanos, PEMEX, the state-owned oil company, for using published data on its industrial activities in the GOM; the Kishon River Authority, Haifa, Israel, for the supply of information in its annual reports on the pollutants in the water of the Kishon river and port; Centro de Investigaciones en Corrosión (CICORR), Universidad Autónoma de Campeche, Mexico, for providing data on corrosion and pollution in the Sound of Campeche, GOM.

21.18 References

4ᵗʰ Ballast Water Technology Conference [Internet]. 2015. UK: IMarEST; [cited 2016 Apr 22]. Available from: http://www.imarest.org/eventscourses/eventsconferences/ ballastwatertechnologyconference/4thballastwatertechnologyconference.

18ᵗʰ ICMCF [Internet]. 2016. IK: IMarEST; [cited 2016 Apr 22]. Available from: http://icmcf2016.univtln.fr/.

Acuña N., Schorr M., Aubert B., Flores J. 2002. Behavior of stainless steel in the Gulf of Mexico seawater. Mater Perform. 41:58–62.

Anonymous. 2013. The future of the oceans: acid test. Economist. Nov: 76–77.

ASTM Standard Annual Book. 1963. Volume 03.02 Wear and Erosion; Metal Corrosion. ASTM International. West Conshohocken, PA.

Bonilla H. R., Mozqueda M. C., Calderon L. E., Diaz G. 2011. La acidificación del océano y los arrecifes del Pacifico Mexicano. Ciencia, Tecnologia e Innovacion para el Desarrollo de Mexico. 3:1.

Cheng N., Cheng J., Valdez B., Schorr M., Bastidas J. M. 2016. Inhibition of seawater steel corrosion via colloid formation. Mater Perform. 55:48–51.

Eliezer A., Valdez B., Schorr M. 2010. Corrosion of naval aluminum in sea water. Mater Perform. 149:62–66.

Hellio C., Yebra D., editors. 2009. Advances in marine antifouling coatings and technologies. Boca Raton: CRC/WP.

Hernandez G., Schorr M., Capio J., Martinez L. 1995. Preservation of the infrastructure in the Gulf of Mexico. Corros Rev. 13:65–80.

Huang K., McLandsborough L. A., Goddard J. M. 2016. Adhesion and removal kinetics of Bacillus cereus biofilms on NiPTFE modified stainless steel. Biofouling. 32:523–533.

Hummel R. 2014. Alternative futures for corrosion and degradation research. Potomac Institute for Policy Studies. Chapter 1, The need for improved corrosion control; p. 58–59.

Leon H. Charney School of Marine Sciences, Maritime Civilization [Internet]. 2016. Israel: University of Haifa; [cited 2016 May 1]. Available from: http://marsci.haifa.ac.il/.

Marcos M., Botana J., Valdez B., Schorr M. 2006. Polución y corrosión en ríos contaminados. Paper presented at: III Congreso de Ingeniería Civil, Territorio y Medio Ambiente; Zaragoza, Spain.

Melcher R. E. 2005. Effect of the nutrient-based water pollution on the corrosion of mild steel in marine immersion conditions. Corros. 61:237–245.

NACE TM 01692000. 2000. Standard guide for laboratory immersion corrosion testing and metals. Houston, TX: NACE.

Raichev R., Veleva L., Valdez B. 2009. Corrosión de metales y degradación de materiales [Metals corrosion and materials degradation]. Mexico: Universidad Autónoma de Baja California.

Railkin A. I. 2004 Marine biofouling: colonization process and defences. Boca Raton: CRC Press.

Roberge P. 2010. Impact of climate change on corrosion risk. In special issue: Valdez B, Schorr M, editors. Relationship of corrosion with climate change. Corros Eng Sci Technol. 45:34–41.

Schorr M., Valdez B. 2005. Corrosion of the marine infrastructure in polluted seaports. Corros Eng Sci Technol. 40:137–142.

Schorr M., Valdez B., Quintero M. 2006. Effect of H_2S on corrosion in polluted waters: a review. Corros Eng Sci Technol. 41:221–227.

Schultz M. P., Bendick J. A., Holm E. R., Hertel W. M. 2011. Economic impact of biofouling on a naval surface ship. Biofouling. 27:87–98.

Valdez B., Schorr M., Bastidas J. M. 2015. The natural gas industry: equipment, materials and corrosion. Corros Rev. 33:175–185.

Valdez B., Schorr M., Quintero M., Garcia R., Rosas N. 2010. Effect of climate change on durability of engineering materials in hydraulic infrastructure: an overview. In special issue: Valdez B., Schorr M., editors. Relationship of corrosion with climate change. Corros Eng Sci Technol. 45(1): 34–41.

Valdez B., Schorr M., Ramos R., Salinas R., Nedev N., Curiel M. 2016. Improved phosphate conversion coating of steel for corrosion protection. Innov Corros Mater Sci. 6:49–54.

Valdez B., Schorr M., So A., Eliezer A. 2011. LNG regasification plants: materials and corrosion. Mater Perform. 50:64–68.

Printed in the United States
by Baker & Taylor Publisher Services